# Randomly Deployed Wireless Sensor Networks

# Randomly Deployed Wireless Sensor Networks

**Xi Chen**
Tsinghua University
Beijing, China

Elsevier
Radarweg 29, PO Box 211, 1000 AE Amsterdam, Netherlands
The Boulevard, Langford Lane, Kidlington, Oxford OX5 1GB, United Kingdom
50 Hampshire Street, 5th Floor, Cambridge, MA 02139, United States

**Notices**

Knowledge and best practice in this field are constantly changing. As new research and experience
broaden our understanding, changes in research methods, professional practices, or medical treatment
may become necessary.

Practitioners and researchers must always rely on their own experience and knowledge in evaluating
and using any information, methods, compounds, or experiments described herein. In using such
information or methods they should be mindful of their own safety and the safety of others, including
parties for whom they have a professional responsibility.

To the fullest extent of the law, neither the Publisher nor the authors, contributors, or editors, assume
any liability for any injury and/or damage to persons or property as a matter of products liability,
negligence or otherwise, or from any use or operation of any methods, products, instructions, or ideas
contained in the material herein.

**Library of Congress Cataloging-in-Publication Data**
A catalog record for this book is available from the Library of Congress

**British Library Cataloguing-in-Publication Data**
A catalogue record for this book is available from the British Library

ISBN: 978-0-12-819624-3

For information on all Elsevier publications
visit our website at https://www.elsevier.com/books-and-journals

*Publisher:* Mara Conner
*Acquisitions Editor:* Glyn Jones
*Editorial Project Manager:* Naomi Robertson
*Production Project Manager:* Poulouse Joseph
*Designer:* Greg Harris

Typeset by VTeX

# Contents

# Acknowledgments

A large portion of the book is based on my work on wireless sensor networks since I joined in the Center for Intelligent and Networked Systems, Department of Automation, Tsinghua University.

I first would like to express my sincere thanks to Prof. Yu-Chi Ho for his advice and help. His insights and inspiration have had a great impact on my research.

I would like to thank Prof. Xiaohong Guan for his leadership and support.

I would also like to thank all colleagues and students who have worked with me and published papers on wireless sensor networks: Hongxing Bai, Junfeng Ge, Dianfei Han, Qingshan Jia, Yongheng Jiang, Bing Li, Jianghai Li, Jie Li, Chunkai Nie, Wen Tang, Xingshi Wang, Yongcai Wang, Li Xia, Ruixi Yuan, Jiansong Zhang, Qianchuan Zhao, and Hongchao Zhou.

My thanks also go to the three anonymous reviewers of the book proposal. I appreciate their helpful comments and suggestions.

I am grateful to those who read drafts of this material and made suggestions to improve it: Guanghong Han, Zixuan Li, Anbang Liu, Renlin Liu, Jinghui Zhang, and Xiwei Zheng. I owe thanks to Elsevier and Tsinghua University Press.

Finally, I would like to thank Tsinghua University, National Natural Science Foundation of China, National Key Research and Development Project of China for providing me with financial support as well as excellent research environment.

This work is partially supported by the National Key Research and Development Project of China (No. 2017YFC0704100) and the National Natural Science Foundation of China (No. 60574064 and 60574087).

Xi Chen
Department of Automation
Tsinghua University
Beijing, China
December 2019

# Introduction

## CONTENTS

Recent advances in wireless communications and micro electromechanical system have enabled the development of low-cost, low-power, multi-functional sensors that are small in size and can communicate with each other from short distances [1]. In the technology review of MIT's Magazine of Innovation in February 2003, Wireless sensor networks (WSNs) come first of the 10 emerging technologies that will change the world.

## 1.1 Overview of WSNs

A WSN usually consists of sensor nodes, sink nodes (or base stations) and end users. Sensor nodes are deployed in the area of interest to form a self-organized network

Randomly Deployed Wireless Sensor Networks. https://doi.org/10.1016/B978-0-12-819624-3.00006-9

to monitor the environments and transmit the phenomenon they sensed to sink nodes with some routing protocols. Sink nodes are often more powerful in computation and long-range communication. As shown in Fig. 1.1, sensor nodes transmit their sensed data to the sink node by multihop relays; the sink node implements data fusion and then sends the information to end users through Internet, satellite or mobile access point (e.g. unmanned aerial vehicle). This information helps users to monitor the environments. Conversely, end users can also send commands to the network for task configuration or information inquiring.

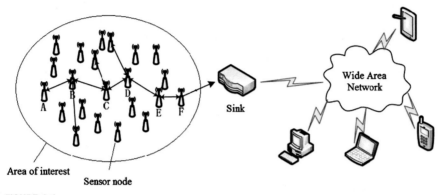

**FIGURE 1.1**

Typical WSN architecture.

As shown in Fig. 1.2, sensor nodes are usually composed of sensing, data processing, communicating and power modules. In addition, depending on the application requirements, sensor nodes may also include a localization system (e.g. GPS), an energy harvesting module (e.g. solar battery), a locomotory module, etc.

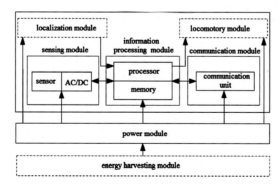

**FIGURE 1.2**

Composition of a sensor node.

Fig. 1.3 presents the photograph of Mica2, which is an early production of Crossbow Technology, Inc. It is powered by two AA batteries. The antenna is 10 centimeters in length and its CPU is ATmega128L, 4K RAM and 128K Flash. The sensor board can integrate various microsensors, such as photosensors, thermal sensors, magnetic sensors, accelerometers, microphones and buzzers.

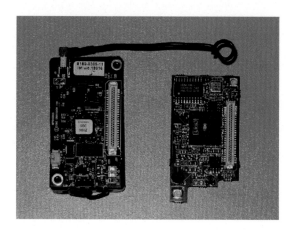

**FIGURE 1.3**

Mica2 mote (left) and Mica2 sensor board (right).

Different from other wireless networks, e.g. mobile communication networks, wireless local area networks and bluetooth networks, WSNs have the following features.

1. Sensor nodes are limited in power, computational capacity and memory. As sensor nodes are usually powered by batteries, energy balancing and energy efficiency are crucial for prolonging the network lifetime. As energy consumption is proportion to the $n$th ($n \geqslant 2$) power of the communication distance, multihop routing is preferable.
2. There are a large number of sensor nodes in a network. More sensor nodes which are deployed in the area of interest can perform better in monitoring the environment. However, more sensor nodes will reduce the network utilization and increase the cost. The key point is the tradeoff between network performance and cost.
3. After being deployed, sensor nodes are self-organized into a network and collaborate to accomplish a common task. There is no strict central node. Malfunction or failure of a single node can hardly affect network performance. Sensor nodes cooperate with a hierarchy protocol and a distributed algorithm.
4. The topology of a network may change with time. Sensor nodes may become invalid as their power is used up or defunct due to random events. Other nodes may join the network to meet the requirement on performance. To extend the network lifetime, working schemes which let each node switch between working and sleeping states also result in a dynamic topology.

## 1.2  Research topics in WSNs

The study on hardware of sensor nodes involves an embedded processing unit, a communication device and sensing elements, while the research on software of sensor nodes focuses on operating systems and programming languages. Besides meeting the requirements on data processing, storage, communication and sensing, both hardware and software should be designed to optimize energy consumption. Some main topics on sensor nodes are as follows.

### Sensing technology

Various applications may demand sensor nodes to collect different physical signals. The design of sensor nodes which are high-integrated, multi-functionalized and miniaturized has been the focus of the study.

### Low-power-consumption sensor nodes

Due to the limits on cost and volume, sensor nodes are usually powered with batteries and cannot be recharged. It is highly necessary to research and develop low-power-consumption sensor nodes which can work much longer.

### Low-cost sensor nodes

A WSN is composed of a large number of sensor nodes. To make WSNs affordable, more efforts need to be made to lower the cost of sensor node while ensuring its performance.

### Wireless communication technology

Sensor nodes need simple, low-cost and robust communication technology. Technologies of carrier, antennae, and data modulation-demodulation are main subjects in the field of sensor nodes.

The research in sensor nodes has been well established from prototype systems to commercial applications. An earlier product series developed by the University of California, Berkeley, and Crossbow Technology, Inc. include Mica, Mica2, Mica2Dot, and MicaZ with operating system TinyOS and programming language Nesc. They have been widely applied in research and development of WSNs.

### Communication protocols

For WSNs, communication protocols usually consist of protocols for the transport layer, the network layer and the data-link layer.

- Transport layer
  Due to the application-oriented and collaborative nature of WSNs, the main data flow occurs in the forward path as the source nodes transmit their data to the sink.

The data originating from the sink, such as programming/retasking binaries, queries, and commands, is sent to the source nodes in the reverse path. Different from traditional communication networks, at the transport layer, the WSN paradigm demands an event-to-sink reliability notion. It is necessary to implement transport layer congestion control in the forward path to ensure reliable event detection at the sink [36].

Event-to-sink reliable transport in WSNs differs from the conventional end-to-end reliable transport in wireless networks. The latter transport is based on acknowledgments and end-to-end retransmissions. These mechanisms for strict end-to-end reliability are unnecessary and spend too much energy. For WSNs, transport layer protocols should achieve reliable event detection with minimizing energy consumption [26].

The flow in the reverse path mainly contains data transmitted by the sink for an operational or application-specific purpose. Dissemination of this data mostly requires 100% reliable delivery. This strict reliability requirement for the sink-to-sensors transport demands retransmission and acknowledgment mechanisms. However, to save energy, local retransmission and negative acknowledgment approaches would be more suitable than end-to-end retransmission and acknowledgments [35].

- Network layer

Energy efficiency is most important for network layer protocols. In WSNs, to integrate tightly with the information or data, a routing protocol may be designed according to data-centric techniques [21,30]. A data-centric routing protocol should apply a data aggregation technique to solve the implosion and overlap problems in routing [11]. Data aggregation combines data from many sensor nodes into a set of meaningful information items [12].

- Data-link layer

In general, the data-link layer is primarily responsible for multiplexing data streams, data frame detection, medium access, and error control; it guarantees reliable point-to-point and point-to-multipoint connections in a communication network. The medium access control (MAC) layer protocols in a multihop WSN need to establish data communication links for multihop wireless communication in a densely scattered sensor field and regulate access to shared media to let sensor nodes fairly and efficiently share communication resources. The MAC protocol must have built-in power conservation, mobility management and failure recovery strategies [35].

There are three kinds of MAC schemes applicable to WSNs. MAC protocols based on time-division multiple access schemes can reduce the duty cycle of the radio. Hence they can prevent collisions and save more energy than contention-based schemes. A pure time-division multiple access-based access scheme dedicates the entire channel to a single sensor node, while a pure frequency-division multiple access scheme allocates a minimum signal bandwidth per node. The tradeoff between the access capacity and the energy consumption plays a key role in MAC schemes. For a hybrid TDMA/FDMA-based access scheme, the optimum number of channels depends on the ratio of the power consumption of the transmitter to that of the receiver. A frequency-division multiple access scheme is better when the transmitter consumes more power. When the receiver consumes a great amount of power, the scheme should lean toward frequency-division multiple access [33]. MAC protocols for WSNs need to support variable but highly correlated and dominantly periodic traffic. The listening mechanism and the backoff scheme are two important components for any carrier sensing medium access-based scheme. The constant listen periods are energy efficient and the introduction of random delay provides robustness against repeated collisions [40].

In WSNs, another important function of the data-link layer is error control of the transmitted data. The main approaches of the error control mechanisms in communication networks are forward error control and automatic repeat request. Since automatic repeat request-based error control mechanism incurs significant additional retransmission cost and overhead, it is not very useful in WSN applications. On the other hand, forward error control schemes have inherent decoding complexity, which requires great processing resources in sensor nodes. Therefore, simple error control codes with low-complexity encoding and decoding may be the best solutions for error control in WSNs.

## Node localization

As sensor nodes may be randomly deployed in the area of interest, they must be aware of their locations in order to provide meaningful data to the users. Location information may also be required by the network and data-link layer protocols. A localization protocol must be robust to node failure, have little sensitivity to measurement noise, be flexible in any terrain and have low error in location estimation. There are two types of localization techniques: beacon based and relative location based.

Beacon-based methods require few nodes to have a well-defined location through GPS or through manual configuration. By ranging and estimation, nodes can discover their location. During the ranging phase, each node estimates the range of its neighbors. And in the estimation phase, nodes whose locations are unknown use the range and the beacons' locations to estimate their locations [27]. In addition, mobile beacon mechanisms provide economical and effective localization signal coverage for sensor nodes [31].

Beacon-based localization protocols are sufficient for certain WSN applications. However, some WSNs may be deployed in areas unreachable by beacons or GPS. To obtain precise relative location information, sensor nodes need to detect the location of neighbors and cooperate with each other for location estimation.

## Time synchronization

Like location information, time information is a crucial element for a fully described event. Sensor nodes in the area of interest must maintain a similar time within a certain tolerance throughout the network lifetime. For short-distance multihop broadcasting, the data processing time and the variation of the data processing times may result in time fluctuations and differences in path delays. In addition, due to the wandering effect of local clocks, the time difference between two sensor nodes may become significant over time. There are three types of timing techniques. The first one relies on fixed time servers, which are expected to be robust and highly precise to synchronize the network. Sensor nodes are synchronized to these time servers. The second one translates time hop-by-hop from the source to the sink throughout the network. And the third one synchronizes the network by automatically organizing and determining the master nodes as the temporary time servers. Different timing techniques and communication protocols may be applied to different applications to meet their requirements on the quality of service of the network [35].

## Network security

In WSNs, the low-cost sensor devices may not be built with high reliability. Moreover, sensor nodes are often deployed in an unattended environment or even hostile circumstance. They communicate with each other using wireless signals which are easy to eavesdrop. There are various kinds of attacks against WSNs. It is important to guarantee that data are safely received by the sink. So far, there are two kinds of security solutions: prevention based and detection based. Typical prevention-based techniques are encryption and authentication. Detection-based techniques are designed to identify and isolate attackers after prevention-based techniques fail.

Research on security mainly focuses on key management, secure routing protocol, denial of service attacks and intrusion detection algorithm. Key management helps to establish secure communications in networks. Self-healing group-wise key schemes with time-limited node revocation guarantee secrecy, and ascertain collusion freedom and group confidentiality in high packet loss environments [32]. For densely deployed WSNs, secure energy-efficient routing protocol can ensure data transmission security and make good use of the available amount of energy in the network [23]. Two general approaches for broadcast authentication in WSNs are digital signature and $\mu$TESLA-based techniques. Both of them are vulnerable to denial of service attacks. By simply forging a large number of broadcast messages with digital signatures, an attacker may force sensor nodes to verify these signatures, and eventually deplete their battery power. By using an efficiently verifiable weak authenticator along with broadcast authentication, a sensor node needs to perform the expensive signature verification or packet forwarding only when the weak authenticator is verified [22]. Due to the tree structure topology of a WSN, path-based denial of service attacks may exhaust the batteries of several nodes and have the potential to disable a much wider region than a single path. A method is proposed to detect path-based denial of service attacks and then set up a cost-effective recovery process by using mobile agents in the network [17]. An efficient distributed group-based intrusion detection algorithm is developed to identify malicious attackers [18]. An overhearing-based detection mechanism is presented to deal with malicious packet-modifying attacks from some malicious nodes which modify contents of data packets while relaying [34].

## Application layer protocols

At the application layer of WSNs, there are mainly three protocols: the sensor management protocol; the task assignment and data advertisement protocol; and the sensor query and data dissemination protocol. System administrators can interact with WSNs with sensor management protocol. Sensor nodes do not have global IDs and usually have no infrastructure supporting. Hence a sensor management protocol needs to access the nodes by using attribute-based naming and location-based addressing. The administrative tasks for a sensor management protocol include sensor nodes' work scheduling, data processing, time synchronization, security in data communication, and network configuration and reconfiguration. Users may be interested in a certain attribute of a phenomenon or a specific event. The task assignment and data advertisement protocol provides a software interface between a WSN and its users. Users may have an interest in a sensor node, a subset of sensor nodes or the whole network, while sensor nodes can advertise the available data to the users. The sensor query and data dissemination protocol provides interfaces to issue queries, respond to queries, and collect incoming replies. These queries are generally attribute-based or location-based naming instead of sending to particular nodes [35].

As the power cannot be replenished for the battery-driven sensor nodes, to extend the network lifetime, energy efficiency is always a critical issue of research in WSNs.

## 1.3 Applications of WSNs

WSNs have been widely applied in military, agriculture, industry, environment monitoring, and various daily life fields. We introduce their applications with a few specific examples.

## Military applications

WSNs are used by the military for a number of purposes such as monitoring militant activity in remote areas and force protection. Being equipped with appropriate sensor nodes, these networks enable detection of enemy movement, identification of enemy force and analysis of their movement and progress [39].

Acoustic sensor arrays suspended below tethered aerostats are used to detect and localize transient signals from mortars, artillery and small arms fire. The airborne acoustic sensor array calculates an azimuth and elevation to the originating transient, and immediately cues a collocated imager. By providing additional solution vectors from several ground-based acoustic arrays, unattended ground sensor systems can augment aerostat arrays to perform a 3D triangulation on a source location [29].

The anti-submarine warfare concept uses small sensors with passive and active sonar to detect modern diesel submarines operating on batteries. Hundreds or thousands of sensor nodes are deployed to provide a high density sensor field. Low-cost sensor nodes have a short detection range. Hence they are far less susceptible to multipath reverberations and other acoustic artifacts [37].

The capability of a WSN military application is dependent on the type and capabilities of sensor nodes, wireless communications architecture, coverage, and appropriate information processing, fusion and knowledge generation. For more military applications, the reader may refer to [7].

## Precision agriculture

WSNs are increasingly applied in agricultural industry. Gravity feed water systems can be monitored by using a wireless network. Pressure transmitters can be used to monitor water tank levels and pumps. Water use can be measured and wirelessly transmitted back to a central control center for billing. Irrigation automation enables more efficient water use and reduced waste [6].

Low production in agriculture is mainly due to the absence of water in the soil. For agricultural monitoring, the measurement of moisture level of plants may be quantified with the difference between the leaf temperature and the air temperature. A novel leaf temperature sensor is developed and the network consisting of such sensors measures the water stress on plants [5].

A WSN composed of 64 sensors is set up to monitor a commercial vineyard. The system provides software modules ranging from filtering raw data to a centralized and a distributed data storage applications. The network provides better geographical coverage and an increased spatial resolution compared to traditional solutions based on individual weather stations [14].

For more agriculture applications, the reader may refer to [25].

## Industrial applications

Industrial WSNs, which incorporate WSNs with intelligent industrial systems, provide many advantages over existing industrial applications, such as low cost, rapid deployment, wireless communication, intelligent controlling, self-organization, and processing capability. Industrial WSNs technologies play an important role in developing more efficient, stable, reliable, flexible, and application-centric industrial systems. In 2014, the International Journal of Distributed Sensor Networks published a special issue on industrial WSNs, which introduces a collection of 17 papers covering a range of topics for specific industrial requirements [19].

Microwireless sensors can be embedded into the key positions of manufacturing equipments or products to monitor them massively in realtime mode. These sensor nodes diagnose and predict faults promptly so that the reliability of the equipment is enhanced and the accident rate is reduced. For example, a WSN is deployed to detect the axle temperature for a freight train [44].

WSNs can also monitor the conditions of industry building facilities and predict any possible problems with data fusion. It is much more convenient and accurate than traditional approaches.

## Environmental monitoring

Space WSNs for Planetary Exploration, as an European Commission collaborative project, present a new approach to using WSNs for planetary surface characterization. Spatially distributed sensors in the proposed networks cooperate to monitor physical and environmental conditions and send their data through a network to a central processing location. Hundreds of small wireless sensors (also called smart dust) may be dropped from an orbiting satellite onto the Moon surface to ensure a uniform and sufficient coverage. These sensor nodes create their own ad hoc network while some of them, which are equipped with satellite communication capabilities, establish a link between the WSN and an orbiter or directly with the Earth. Data gathered from the sensors is processed and sent to the orbiter and later to the Earth. The sensor for this project is a micro-meteorological station, which can measure temperature, radiation, dust deposition and irradiance at different wavelengths. Each station is autonomously powered and has networking and data processing capabilities [43].

WSNs can greatly assist the geophysics community. Studying active volcanoes demands high data rates, high data fidelity, and sparse arrays with high spatial separation between nodes. A WSN is designed and deployed on Volcan Reventador, which is located in northern Ecuador, for volcanic data collection and reliable data retrieval [38]. WSNs are also used to monitor snow composition in mountain slopes for avalanche forecasting [2].

The Great Duck Island project is a good example of habitat and microclimate monitoring. A WSN is used for the study of Leach's storm petrel on Great Duck Island, Maine, USA. The petrel-watching apparatus consists of a distributed system of devices called 'motes'. The mote contains application-specific sensors and signal-processing hardware for data collection and has a low-power radio transceiver for communication. When the motes are networked together, each simultaneously collects data from its immediate surroundings and passes its own and other motes' data through the network [16].

Forest fires are a threat to sustainability of the forest. As surveillance systems for forest fires, WSNs gather data values, like temperature and humidity, from all points of an area incessantly, day and night, and relay fresh and precise data to the fire-fighting center quickly [42].

## Health monitoring

WSNs in healthcare can improve the way of patients being monitored in an infirmary or a hospital. The healthcare WSNs collect and send patients' health parameters, such as blood pressure, pulse, body temperature, wirelessly to remote monitoring systems. The ability to let patients moving around when hospitalized is important to promote their quality of life [4]. With multihop network communication, compared to classical wired or large inspection equipment, micro-sensors can be used in long-term monitoring without causing any inconvenience to patients. WSNs can provide more convenient and swift solutions in remote telemedicine, hospital medicine management and intelligent home nursing for the aged.

## Smart city

Urban issues are becoming more complicated and complex as the city grows. Implementing the concept of a smart city is one solution to better management, which is helpful to create a comfortable, safe, and sustainable city atmosphere [20]. In many countries around the world, smart cities are becoming a reality. These cities make efforts to improve the citizens' quality of life by providing services that are normally based on data extracted from WSNs and other elements of the Internet of Things. In addition, public administration uses these smart city data to increase its efficiency, to reduce costs and to provide additional services [10]. Wireless networks will be one of the key technologies for road traffic management in smart cities. Vehicles and dedicated roadside units should be interconnected through wireless technologies. Traffic lights and road signs form a large-scale network of small devices that report measurements, take orders from a control center, and are able to make decisions autonomously based on their local perception [8].

There is more work on smart cities. The reader may refer to [3,24,41].

## Civil engineering

WSNs have demonstrated their potential for providing continuous structural response data to quantitatively assess structural health and have become a practical tool for structural health monitoring of large, complex civil structures. Monitoring and analyzing the health of large structures like buildings, bridges, dams, and heavy machinery is important for safety, cost, operation, making prior protective measures, and repair and maintenance. With WSN composed of a MEMS accelerometer, health monitoring can be performed by studying the dynamic response through measuring of ambient vibrations and strong motion of such structures [15].

Blockages in sewers are major causes of sewer flooding and pollution. The detection of the sewer condition is necessary to prevent flooding. A WSN composed of a large number of diverse sensor nodes distributed within a sewer infrastructure network can detect blockages proactively, and then feed these event data back to a central control room [28].

## Other applications

Radio Frequency IDentification (RFID) is an important wireless technology which has a wide variety of applications and provides unlimited potential. RFID is used to detect the presence and location of objects, while a WSN is used to sense and monitor the environment. Integrating RFID with WSN not only provides identity and location of an object but also provides information regarding the condition of the object which carries the sensors enabling RFID tag. It can be widely used in military, environmental monitoring and forecasting, warehouse management, healthcare, intelligent transport vehicles, intelligent home, and precision agriculture [13].

The Internet of Things (IoT) represents the network of physical sensors which are able to sense changes in the environment, communicate with each other, and send data to the external environment. WSNs play a critical role in IoT. It is essentially the interface between the IoT and the physical world [9].

## 1.4 **Outline of the book**

This book aims to introduce our research on randomly deployed WSNs. It consists of seven chapters. This chapter introduces WSNs and presents some research topics and applications of WSNs. The rest of the book is organized as follows.

For a randomly deployed WSN, coverage performance is essentially dependent on each sensor node's position so that complete coverage is too demanding to be easily achievable. Chapter 2 considers point coverage as an alternative measure of coverage performance to complete coverage. Point coverage probabilities are derived. The relationship between the number of sensors and coverage probability is analyzed. Comparisons between complete coverage and point coverage are made by analysis and simulation. Performance optimization and cost control are discussed for heterogeneous WSNs.

Chapter 3 presents a location-based percentage coverage configuration protocol and a localized location-free node scheduling algorithm. Both of them can secure a desired percentage of the area of interest being covered. In many applications, a small loss of coverage is acceptable and can save much energy.

Chapter 4 studies the design and configuration of a WSN for target detection. The problem is to let the network secure a certain detection probability and work for a reasonable long time. One novel model of randomly distributed network is presented. A state switching scheme is developed and a two-level optimization problem is formulated to solve the problem.

Chapter 5 presents a novel probabilistic forwarding approach for directed data transmission without route discovery. In this model, each message is required to reach the base station successfully with a predefined probability, hence sensor nodes which are located nearer to the base station need to relay messages with a certain relay probability. The relationship between the number of relaying nodes and relay probability is analyzed and the condition for relay probability to guarantee the success probability is obtained.

Chapter 6 introduces a stochastic scheduling mechanism to meet the requirement on coverage performance and prolong the network lifetime. An algorithm is developed to determine the working probability based on the number of effective nodes in each period. The impact of failure nodes on coverage performance is also studied. Moreover, a revision on dynamic working probability is presented to compensate the loss of working nodes. An algorithm is proposed to optimize the period length to maximize the network lifetime.

Chapter 7 considers an energy-based acoustic multisource localization problem. When the source number is given, a multisource localization problem can be formulated as a maximum likelihood estimation problem. A simple and efficient localization method which combines location initialization with multiresolution search is proposed. Moreover, two methods based on node clustering and the minimum description length criterion are separately developed to estimate the unknown source number.

# References

[1] I.F. Akyildiz, W. Su, Y. Sankarasubramaniam, E. Cayirci, A survey on sensor networks, IEEE Communications Magazine (August 2002) 102–114.

[2] C. Alippi, G. Anastasi, C. Galperti, F. Mancini, M. Roveri, Adaptive sampling for energy conservation in wireless sensor networks for snow monitoring applications, in: IEEE International Workshop on Mobile Ad Hoc and Sensor Systems for Global and Homeland Security, MASS-GHS, Pisa, Italy, October 8, 2007.

[3] K. Antonopoulos, C. Petropoulos, C.P. Antonopoulos, N. Voros, Security data management process and its impact on smart cities' wireless sensor networks, in: South Eastern European Design Automation, Computer Engineering, Computer Networks and Social Media Conference, SEEDA-CECNSM, 2017.

[4] J.M.L.P. Caldeira, J. Rodrigues, P. Lorenz, Intra-mobility support solutions for healthcare wireless sensor networks-handover issues, IEEE Sensors Journal 13 (11) (2013) 4339–4348.

[5] S.N. Daskalakis, A. Ana Collado, A. Apostolos Georgiadis, M.M. Tentzeris, Backscatter Morse leaf sensor for agricultural wireless sensor networks, in: Conference: IEEE SENSORS, Glasgow, Scotland, UK, October 2017.

[6] S. Dhamdhere, S.V. Gumaste, Monitoring applications in wireless sensor networks, International Journal of Advance Foundation and Research in Computer (IJAFRC) 2 (6) (June 2015) 27–30.

[7] M.P. Durisic, Z. Tafa, G. Goran Dimic, V. Milutinovic, A survey of military applications of wireless sensor networks, in: 2012 Mediterranean Conference on Embedded Computing, MECO, IEEE, 2012.

[8] S. Faye, C. Chaudet, Characterizing the topology of an urban wireless sensor network for road traffic management, IEEE Transactions on Vehicular Technology 65 (7) (2016) 5720–5725.

[9] D. Fong, Wireless sensor networks, in: Hwaiyu Geng (Ed.), Internet of Things and Data Analytics Handbook, Wiley, 2017, Chapter 12.

[10] V. Garcia-Font, C. Garrigues, H.R. Pous, A comparative study of anomaly detection techniques for smart city wireless sensor networks, Sensors 16 (6) (2016) 868.

[11] W.R. Heinzelman, J. Kulik, H. Balakrishnan, Adaptive protocols for information dissemination in wireless sensor networks, in: ACM Mobicom'99, Seattle, WA, August 1999, pp. 174–185.

[12] W.R. Heinzelman, A. Chandrakasan, H. Balakrishnan, Energy-efficient communication protocol for wireless microsensor networks, in: IEEE Proc. Hawaii Int. Conf. Syst. Sci., Maui, HI, January 2000, pp. 1–10.

[13] P.C. Jain, K.P. Vijaygopalan, RFID and wireless sensor network, http://www.scribd.com/doc/32593709/RFID-and-Wireless-Sensor-Network, last retrieved, December 1, 2010.

[14] C. Jardak, K. Rerkrai, A. Kovacevic, J. Riihijarvi, P. Mahonen, Design of large-scale agricultural wireless sensor networks: email from the vineyard, International Journal of Sensor Networks 8 (2) (2010) 77–88.

[15] T. Kijewski-Correa, M. Haenggi, P. Antsaklis, Wireless sensor networks for structural health monitoring: a multi-scale approach, in: 2006 ASCE Structures Congress, 17th Analysis and Computation Specialty Conference, St. Louis, MO, May 18–21, 2006.

[16] J. Kumagai, Life of birds [wireless sensor network for bird study], IEEE Spectrum 41 (4) (2004) 42–49.

[17] B. Li, L. Batten, Using mobile agents to recover from node and database compromise in path-based DoS attacks in wireless sensor networks, Journal of Network and Computer Applications 32 (2) (2009) 377–387.

[18] G.R. Li, J.S. He, Y.F. Fu, Group-based intrusion detection system in wireless sensor networks, Computer Communications 31 (18) (2008) 4324–4332.

[19] S. Li, et al., Industrial wireless sensor networks, International Journal of Distributed Sensor Networks (August 2014) 1–2.

[20] E.A. Lisangan, S.C. Sumarta, Proposed prototype and simulation of wireless smart city: wireless sensor network for congestion and flood detection in Makassar, in: The 2nd East Indonesia Conference on Computer and Information Technology, Makassar, Indonesia, November 6–7, 2018.

[21] J. Mirkovic, et al., A self-organizing approach to data forwarding in large-scale sensor networks, in: IEEE Int. Conf. Commun. ICC'01, Helsinki, Finland, June 2001.

[22] P. Ning, A. Liu, W. Du, Mitigating DoS attacks against broadcast authentication in wireless sensor networks, ACM Transactions on Sensor Networks 4 (1) (2008) 1–35.

[23] A.S. Pathan, C.S. Hong, SERP: secure energy-efficient routing protocol for densely deployed wireless sensor networks, Annals of Telecommunications 63 (9) (2009) 529–541.

[24] J.G. Rojas, L. Camargo, R. Montero, Mobile wireless sensor networks in a smart city, International Journal on Smart Sensing and Intelligent Systems 11 (1) (2018) 1–8.

[25] Romanov, I. Galelyuka, Y. Sarakhan, Wireless sensor networks in agriculture, in: IEEE Seventh International Conference on Intelligent Computing and Information Systems, 2015.

[26] Y. Sankarasubramaniam, O.B. Akan, I.F. Akyildiz, ESRT: event-to-sink reliable transport for wireless sensor networks, in: Proc. ACM MOBIHOC, Annapolis, MD, June 2003, pp. 177–188.

[27] A. Savvides, C. Han, M. Srivastava, Dynamic fine-grained localization in ad hoc networks of sensors, in: Proc. ACM Mobicom'01, Rome, July 2001, pp. 166–179.

[28] C.H. See, et al., A Zigbee based wireless sensor network for sewerage monitoring, in: Asia Pacific Microwave Conference, 2009, pp. 731–734.

[29] M.V. Scanlon, C.G. Reiff, L. Solomon, Aerostat acoustic payload for transient and helicopter detection, in: SPIE Defense & Security Symposium, Orlando, Florida, USA, 2007.

[30] C. Shen, C. Srisathapornphat, C. Jaikaeo, Sensor information networking architecture and applications, IEEE Personal Communications (August 2001) 52–59.

[31] Q. Shi, H. Huo, T. Fang, D. Li, A 3D node localization scheme for wireless sensor networks, IEICE Electronics Express 6 (3) (2009) 167–172.

[32] M. Shi, X. Shen, Y. Jiang, C. Lin, Self-healing group-wise key distribution schemes with time-limed node revocation for wireless sensor networks, Wireless Communications 14 (5) (2007) 38–46.

[33] E. Shih, et al., Physical layer driven protocol and algorithm design for energy-efficient wireless sensor networks, in: ACM Mobicom'01, Rome, July 2001, pp. 271–286.

[34] K.F. Ssu, P.H. Chou, L.W. Cheng, Using overhearing technique to detect malicious packet-modifying attacks in wireless sensor networks, Computer Communications 30 (11) (2007) 2342–2352.

[35] W. Su, E. Cayirci, B.Ö. Akan, Overview of communication protocols for sensor networks, in: I. Maoub, M. Ilyas (Eds.), 4th Part of Sensor Network Protocols, CRC Press, Taylor & Francis Group, 2006.

[36] S. Tilak, N.B. Abu-Ghazaleh, D. Estrin, Directed diffusion: a scalable and robust communication paradigm for sensor networks, in: ACM Mobicom'00, Boston, MA, August 2000, pp. 56–67.

[37] J.P. Towle, R. Johnson, H.T. Vincent II, Low-cost acoustic sensors for littoral anti-submarine warfare (ASW), in: SPIE, Orlando, USA, 2007, pp. 653–814.

[38] G. Werner-Allen, K. Lorincz, M. Ruiz, O. Marcillo, J. Johnson, J. Lees, M. Welsh, Deploying a wireless sensor network on an active volcano, in: IEEE Internet Computing. Special Issue on Data-Driven Applications in Sensor Networks, March/April, 2006.

[39] M. Winkler, K.D. Tuchs, K. Hughes, G. Barclay, Theoretical and practical aspects of military wireless sensor networks, Journal of Telecommunications & Information Technology (2008) 37–45.

[40] A. Woo, D. Culler, A transmission control scheme for media access in sensor networks, in: ACM Mobicom'01, Rome, July 2001, pp. 221–235.

[41] M. Xie, Y. Bai, Z. Hu, C. Shen, Weight-aware sensor deployment in wireless sensor networks for smart cities, Wireless Communications and Mobile Computing (14) (2018) 1–15.

[42] S. Yassine, N. El Kamoun, Forest Fire Detection and Localization With Wireless Sensor Networks, Springer, Berlin, Heidelberg, 2013.

[43] X. Zhai, H. Jing, T. Vladimirova, Multi-sensor data fusion in wireless sensor networks for planetary exploration, in: 2014 NASA/ESA Conference on Adaptive Hardware and Systems, July 2014.

[44] S. Zhang, et al., Design of axle-temperature detect system based on wireless sensor network (in Chinese), Modern Electronics Technique (3) (2008) 86–88.

# Point coverage analysis

## CONTENTS

A wireless sensor network (WSN) is usually deployed in a region to monitor the environment or targets. Such a region may be called the area of interest. Coverage is one important criterion to evaluate the network performance. The coverage problem is how to make the network meet the requirements on coverage performance and prolong its lifetime as well.

## 2.1 Coverage in WSNs: elements

### 2.1.1 Sensor sensing models

The sensing range of an individual sensor node refers to the region it can monitor. The sensing region of a WSN is the union set of all nodes' sensing ranges.

There are various sensor sensing models. For example, in the formulations of exposure-based coverage [5,13,15,17] and information coverage [1,19,20], a sensor node can detect the signal at any distance but its sensing ability diminishes as distance increases. Boolean sensing models and probabilistic sensing models are widely adopted in the study of WSNs.

**Randomly Deployed Wireless Sensor Networks. https://doi.org/10.1016/B978-0-12-819624-3.00007-0**

The Boolean sensing model is also known as the 0-1 sensing model or binary sensing model. In a 2D plane, a node's sensing range is a disc with sensing radius $r$ and centered at the node itself; the value of $r$ is determined by the node's physical properties. One sensor node can observe environments and events which occur in its sensing range. It cannot detect anything outside the range. Similarly, in a 3D plane, the sensing range in a boolean sensing model is a sphere with sensing radius.

Compared to the boolean sensing model, the probabilistic sensing model is more realistic. Instead of being unchanged as assumed in boolean model, a node's sensing ability is likely affected by external environmental factors and diminishes as distance increases. If the probability that node $o$ senses point $a$ is denoted as $v(o,a)$, then $v(o,a)$ may exponentially decrease with the distance between $o$ and $a$, which is denoted as $d(o,a)$. Hence the probabilistic sensing model may be formulated as

$$v(o,a) = \begin{cases} 1, & 0 \le d(o,a) < r - \varepsilon; \\ e^{-\lambda d(o,a)^{\beta}}, & r - \varepsilon \le d(o,a) < r + \varepsilon; \\ 0, & d(o,a) \ge r + \varepsilon, \end{cases}$$

where $\varepsilon > 0$ is a small number, parameters $\lambda > 0$ and $\beta > 0$.

Some other sensing models have been proposed, for example, directional sensing model for video sensors [12] and a polygon model for directional sensors [21].

The Boolean sensing model is an ideal approximation model. It is applicable to the analysis of the network features, and therefore most coverage configuration schemes are based on this model.

## 2.1.2 Coverage formulations

In WSNs, there are several coverage formulations such as complete coverage, point coverage, barrier coverage, etc.

Complete coverage means that a sensor network can sense the whole area of interest without any vacancy (or hole). Complete coverage is ideal and reliable in many situations in which the security (of personnel and/or articles) is of the highest priority. Most studies on complete coverage are based on sensor positions and network topologies.

However, partial coverage may be good enough in some applications. For example, if we wish to deploy a sensor network to monitor a mountain fire, it is neither necessary nor possible to have the mountainous region being completely covered. Point coverage means that a part or a set of points in the area of interest is covered by working sensor nodes. Compared to complete coverage, the performance of point coverage is weaker, but in a randomly deployed WSN, point coverage needs much less working nodes and hence prolongs the network lifetime.

Barrier coverage [3,9] is implemented by laying barriers of sensor nodes in a belt-region to detect intruders.

### 2.1.3 Deployment approaches

There are two approaches for sensor deployment: deterministic deployment and random deployment. Deterministic schemes can let senor nodes be placed at predetermined positions to meet all requirements, such as coverage, connectivity, expected lifetime. However, deterministic schemes are only suitable to controllable environments, e.g. office buildings, hospitals, factories. In harsh or hostile environments, e.g. forests, deserts, battlefields, sensor nodes may be air-dropped from an aircraft or be distributed in other ways, which can result in a random placement [6]. Random schemes usually deploy more nodes and need more complicated sensor scheduling mechanisms.

Some kinds of sensor nodes have mobility. Hence deployments are also classified as static deployment and mobile deployment. In the literature, most studies and applications are based on static deployment. Mobile sensor nodes make the deployment easier and they are more capable of monitoring and tracking moving targets [2,8,11].

### 2.1.4 Scheduling mechanisms

In general, sensor nodes can be in working (or ON) or sleeping (or OFF) state. Sleeping spends less energy. Of course, keeping the node in working state is the simplest scheduling for sensor nodes if they have enough energy or have power supply at any time.

Due to the power limit, sensor nodes usually need to work in rounds and take their turn working to meet the requirement on coverage performance for a certain period of time. Each round consists of configuration and working phases, as shown in Fig. 2.1. In the configuration phase, all sensor nodes follow predefined rules to decide their states in the working phase. At the end of the configuration phase, nodes which are not elected to work will switch OFF and stay sleeping till the end of the round. This mechanism helps to balance the energy consumption among nodes to prolong the network lifetime. But it requires time synchronization for all sensor nodes.

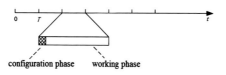

**FIGURE 2.1**

Sensor node's working in rounds mode.

Another scheduling mechanism is defined as follows: in each round, every node will work with a certain probability independently. Sensor nodes need neither negotiation nor time synchronization. This mechanism is much simpler, more energy-efficient, and more suitable to random deployed WSNs.

## 2.2 Point coverage in homogeneous WSNs

Complete coverage and point coverage are different coverage measures. In general, complete $k$-coverage means that any point in the area of interest is covered by at least $k$ ($k \geq 1$) nodes [7]. Complete coverage sounds, of course, more reliable than any kind of partial coverage. However, with random deployment, complete coverage cannot be obtained with certainty but only with a certain probability.

Basically, point coverage is one kind of partial coverage. If one point is covered by at least $k$ sensor nodes, then the point is called point $k$-covered. In general, point $k$-coverage concerns a portion of the area of interest in which any point is $k$-covered. Similar to complete coverage, when sensor nodes are randomly deployed, any point in the area of interest cannot guarantee to be $k$-covered either.

Complete coverage has been well studied. The theory of coverage process provides a basic method for analyzing complete coverage in sensor networks [7]. Using the same decomposition method as in [7], Zhang and Hou [22] analyze the required node density for an asymptotical complete $k$-coverage of the area of interest. They also obtain the upper bound of the lifetime when only a portion of the area of interest is covered. Kumar, Lai, and Balogh [10] consider complete coverage with three kinds of deployments in WSNs: grid deployment, random uniform deployment, and Poisson deployment. In addition, they assume that sensor nodes are working with a probability. Then they derive the conditions for an asymptotic complete coverage of the area of interest under the three deployments, respectively. Wan and Yi [18] assume that sensor nodes are deployed as either a Poisson point process or a uniform point process in a square or disc region. They consider the complicated boundary effect and derive the asymptotic requirement of node density for complete $k$-coverage.

In the rest of this chapter, we consider the coverage problem in an area of interest in 2D plane and impose the following assumptions [4].

**Assumption 2.1.** *The area of interest is a square with side length L (its area is $S = L^2$).*

**Assumption 2.2.** *Sensor nodes are randomly and uniformly distributed in the area of interest.*

**Assumption 2.3.** *The Boolean sensing model is adopted.*

**Assumption 2.4.** *The sensing range of each sensor node is much smaller than S.*

Assumption 2.2 means that all senor nodes are identically and independently distributed in the area of interest following a uniform distribution (called constant diffusion in [16]); in other words, each sensor node falls at any point of the area of interest with identical probability. Due to this assumption, the number of nodes in a portion of the area of interest follows the binomial distribution. In the literature, many studies assume the number of nodes following Poisson distribution (see e.g. [7,22]). However, the fact is that a Poisson distribution is an asymptotic approximation of a binomial distribution.

With Assumption 2.3, the sensing range of a sensor node is $s_r = \pi r^2$ where $r$ is the sensing radius. Moreover, when a sensor node is close to the boundary of the area of interest, part of its sensing range is outside the area of interest. Such a boundary effect could be ignored with Assumption 2.4.

In this section, we first consider a homogeneous WSN and derive the formula for the point $k$-coverage probability and compare point coverage with complete coverage.

### 2.2.1 Point coverage probability

Based on Assumptions 2.2–2.4, we know that each point in the area of interest has the same probability to be $k$-covered. Hence the probability is the point $k$-coverage probability of the network.

**Proposition 2.1.** *Denote by $\alpha_k$ the point k-coverage probability, then*

$$\alpha_k = 1 - \sum_{i=0}^{k-1} \binom{N}{i} (\frac{s_r}{S})^i (1 - \frac{s_r}{S})^{N-i}, \tag{2.1}$$

*where N is the total number of (working) nodes in the network.*

*Proof.* First, $\alpha_1$ is the probability that any point in the area of interest is covered by one or more sensor nodes. If a point is 1-covered, then the point falls in at least one node's sensing range. The probability for the point falling in one node's sensing area is $s_r/S$. The probability that the point is covered by none of the $N$ nodes is $(1 - s_r/S)^N$. Therefore, the point 1-coverage probability is

$$\alpha_1 = 1 - (1 - \frac{s_r}{S})^N. \tag{2.2}$$

Moreover, that a point is $k$-covered means there exist at least $k$ nodes located within the disc of radius $r$ centered at the point. Since nodes are randomly and uniformly deployed in the area of interest, the probability that there are exactly $i$ nodes falling in this disc is

$$\binom{N}{i} (\frac{s_r}{S})^i (1 - \frac{s_r}{S})^{N-i}.$$

Therefore, the probability that there are less than $k$ nodes in the disc is

$$\sum_{i=0}^{k-1} \binom{N}{i} (\frac{s_r}{S})^i (1 - \frac{s_r}{S})^{N-i}.$$

And the probability that there are at least $k$ nodes in the disc is

$$1 - \sum_{i=0}^{k-1} \binom{N}{i} (\frac{s_r}{S})^i (1 - \frac{s_r}{S})^{N-i} = \alpha_k. \qquad \square$$

**Example 2.1** (Forest fire detection). Fire detection is a typical problem of coverage. If a forest is completely covered by a sensor network, any spark or fire in the forest will be detected immediately. On the other hand, if $P$ ($P < 1$) is the portion of the forest that is covered, on average, the detection probability will be $P$. In case a fire happens to start in a coverage vacancy, the fire will be eventually detected as it spreads but may result in a little delay of the fire rescue service.

Suppose the area of a forest is $S_F$; each sensor node has sensing radius $r$ and hence its sensing range is $s_r = \pi r^2$. We require that the sensor network can immediate detect any fire with probability $P^*$.

Considering 1-coverage, if sensor nodes keep working, the number of nodes, $N_F$, should satisfy $P^* = 1 - (1 - s_r/S_F)^{N_F}$, that is,

$$N_F = \frac{\ln(1 - P^*)}{\ln(1 - \frac{s_r}{S_F})}. \tag{2.3}$$

Moreover, the following proposition provides the point $k$-coverage probability as each sensor node works with a certain probability.

**Proposition 2.2.** *Suppose that, at any time, each sensor node works with probability $\rho$. Denote by $\alpha_k(\rho)$ the corresponding point $k$-coverage probability, then*

$$\alpha_k(\rho) = 1 - \sum_{i=0}^{k-1} \binom{N}{i} \left(\frac{\rho s_r}{S}\right)^i \left(1 - \frac{\rho s_r}{S}\right)^{N-i}, \tag{2.4}$$

*where $N$ is the total number of nodes in the network.*

*Proof.* The probability that there are exactly $n$ sensor nodes working in the network is

$$\text{Prob(there are exactly } n \text{ nodes working in the network)} = \binom{N}{n} \rho^n (1 - \rho)^{N-n}.$$

Since each node's position follows a uniform distribution, we have

$$\text{Prob(within the area of } s_r, i \text{ of } n \text{ nodes are working)}$$

$$= \sum_{n=i}^{N} \binom{N}{n} \rho^n (1 - \rho)^{N-n} \binom{n}{i} \left(\frac{s_r}{S}\right)^i \left(1 - \frac{s_r}{S}\right)^{n-i}$$

$$= \sum_{n=i}^{N} \binom{N}{n} \binom{n}{i} \left(\frac{\rho s_r}{S}\right)^i \left(\rho - \frac{\rho s_r}{S}\right)^{n-i} (1 - \rho)^{N-n}$$

setting $l = n - i$

$$= \sum_{l=0}^{N-i} \binom{N}{l+i} \binom{l+i}{i} \left(\frac{\rho s_r}{S}\right)^i \left(\rho - \frac{\rho s_r}{S}\right)^l (1 - \rho)^{N-i-l}$$

$$\text{using} \quad \binom{N}{l+i}\binom{l+i}{i} = \binom{N}{i}\binom{N-i}{l}$$

$$= \sum_{l=0}^{N-i}\binom{N}{i}\binom{N-i}{l}(\frac{\rho s_r}{S})^i(\rho - \frac{\rho s_r}{S})^l(1-\rho)^{N-i-l}$$

$$= \binom{N}{i}(\frac{\rho s_r}{S})^i\sum_{l=0}^{N-i}\binom{N-i}{l}(\rho - \frac{\rho s_r}{S})^l(1-\rho)^{N-i-l}$$

$$= \binom{N}{i}(\frac{\rho s_r}{S})^i(1-\frac{\rho s_r}{S})^{N-i}.$$

Therefore, the point $k$-coverage probability is

$$
\begin{aligned}
\alpha_k(\rho) \quad &= 1 - \sum_{i=0}^{k-1}\text{Prob(within the area of } s_r, i \text{ of } n \text{ nodes are working)}\\
&= 1 - \sum_{i=0}^{k-1}\binom{N}{i}(\frac{\rho s_r}{S})^i(1-\frac{\rho s_r}{S})^{N-i}. \quad \Box
\end{aligned}
\tag{2.5}
$$

As $\rho = 1$, all sensor nodes in the network are working at any time, Eq. (2.4) reduces to Eq. (2.1). In fact, by replacing $s_r/S$ in Eq. (2.1) with $\rho s_r/S$, we obtain Eq. (2.4). $\rho s_r/S$ is the probability that a point is falling in one node's sensing area and the node is working; therefore, it is the probability that the node senses the point.

If there are $N$ nodes in total and every node works with probability $\rho$, then on average there are $\rho N$ nodes working at any time. Suppose every node has total energy $E_o$ and consumes $Q_{on}$ per time unit when working. Hence every node is able to work continuously for $L_o = E_o/Q_{on}$. At any time, on average the energy consumption per unit time for the network is $\rho N Q_{on}$. If the energy spent on state switching is ignored, the network system works for

$$\hat{L}_{sn} = \frac{NE_o}{\rho N Q_{on}} = \frac{E_o}{\rho Q_{on}} = \frac{L_o}{\rho}. \tag{2.6}$$

$\hat{L}_{sn}$ is inversely proportional to nodes' working probability $\rho$. A smaller $\rho$ makes the network work longer. However, smaller $\rho$ results in worse performance. Hence, to guarantee the performance, the network needs more sensor nodes. The reader may refer to Chapter 6 for analysis in detail.

**Example 2.2** (Continuation of Example 2.1). Suppose that each node is able to work for $L_o$ time units but the network is required to normally operate for $k(k > 1)$ times of $L_o$. Hence, the network needs more nodes. Let each node work with probability $\rho = 1/k$ so that the network lifetime becomes $kL_o$. In this way, the average probability for each node to detect a fire becomes $\rho s_r/S_F$. Therefore the number of nodes required by the network is

$$N_{\mathrm{F}} = \frac{\ln(1 - P^*)}{\ln(1 - \frac{\rho s_r}{S_{\mathrm{F}}})} = \frac{\ln(1 - P^*)}{\ln(1 - \frac{s_r}{k S_{\mathrm{F}}})}. \tag{2.7}$$

## 2.2.2 Complete coverage vs. point coverage

Denote by $\omega_k(\rho)$ (or $\omega_k$ when $\rho = 1$) the complete $k$-coverage probability. The accurate value of $\omega_k(\rho)$ has not been obtained yet but its upper bound, denoted as $\hat{\omega}_k(\rho)$, is as follows (see [14]):

$$\hat{\omega}_k(\rho) = \frac{4(k+1)!(N\rho)^{-1}(\frac{N\rho s_r}{S})^{-k} e^{\frac{N\rho s_r}{S}}}{1 + 4(k+1)!(N\rho)^{-1}(\frac{N\rho s_r}{S})^{-k} e^{\frac{N\rho s_r}{S}}}. \tag{2.8}$$

The meaning of $\alpha_k = 0.99$ is that, on the average, 1% of the area of interest will be not $k$-covered. $\alpha_k < 1$ means that there exists $k$-vacancy. When the network has a large number of nodes, even with random deployment, any point in the area of interest is likely to be covered.

Different from the meaning of $\alpha_k = 0.99$, $\omega_k = 0.99$ means that, if we perform 100 independent experiments of randomly and uniformly distributing the $N$ nodes in the area of interest, on average, we will find in 99 experiments that any point in the area of interest is $k$-covered.

Obviously, for the realization of random deployment, the probability of any point to be $k$-covered is greater than the probability that the whole area is $k$-covered, namely $\alpha_k > \omega_k$. According to Eq. (2.8) and Eq. (2.2), $\hat{\omega}_1$ and $\alpha_1$ change with $N$, as shown in Fig. 2.2.

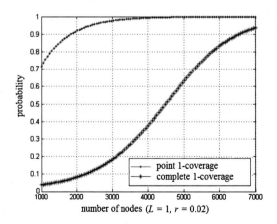

**FIGURE 2.2**

Coverage probability.

Fig. 2.2 shows that the point coverage probability $\alpha_1$ is much bigger than the upper bound of the complete coverage probability $\hat{\omega}_1$. When $N = 1832$, $\alpha_1$ already reaches 0.9, while, when $N = 6534$, $\omega_1$ gets a value of 0.9. In other words, to have the area of interest being point 1-covered with probability 0.9, the WSN only needs 1832 sensor nodes. However, to have the area of interest being completely 1-covered with probability 0.9, the WSN needs 6534 sensor nodes. The number of nodes is more than three times that for point 1-coverage.

Even with a large number of sensor nodes, the network hardly guarantees complete coverage. Complete coverage is not an economical option. In the design and planning of a WSN, if the requirement on coverage performance could be relaxed from complete coverage to point coverage, the cost on sensor nodes will be greatly reduced.

### 2.2.3 Analysis of boundary effect

Due to Assumption 2.4, the boundary effect could be ignored even if the nodes are close to the boundary. However, when the sensing radius is large, we have to take it into account.

Based on Assumption 2.1, we divide the area of interest into a central region and a boundary region, as shown in Fig. 2.3. The central region is the inner square and the boundary region is the one in shadow.

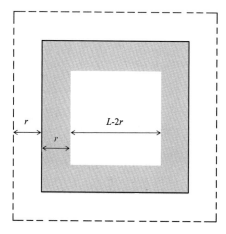

**FIGURE 2.3**

Central region, boundary region and expanded area of interest.

For every point in the central region, the disc centered at the point and with radius $r$ completely falls in the area of interest. Hence the central region is free of the boundary effect. However, any point in the boundary region is more or less affected by the boundary effect and its coverage probability is smaller than that in the central region. Hence Eq. (2.1) or Eq. (2.5) yields the upper bound of the point $k$-coverage probability, namely

$$\hat{\alpha}_k = 1 - \sum_{i=0}^{k-1} \binom{N}{i} (\frac{s_r}{S})^i (1 - \frac{s_r}{S})^{N-i}, \ k \geqslant 1, \tag{2.9}$$

or

$$\hat{\alpha}_k(\rho) = 1 - \sum_{i=0}^{k-1} \binom{N}{i} (\frac{\rho s_r}{S})^i (1 - \frac{\rho s_r}{S})^{N-i}, \ k \geqslant 1. \tag{2.10}$$

Let the area of interest (the middle square with solid line) expand to the outer square with broken line as shown in Fig. 2.3. Its area is $S_x = (L + 2r)^2$. If $N$ sensor nodes are deployed randomly and uniformly in the outer square, then the probability of any point in the middle square to be $k$-covered is

$$\check{\alpha}_k = 1 - \sum_{i=0}^{k-1} \binom{N}{i} (\frac{s_r}{S_x})^i (1 - \frac{s_r}{S_x})^{N-i}, \ k \geqslant 1, \tag{2.11}$$

or

$$\check{\alpha}_k(\rho) = 1 - \sum_{i=0}^{k-1} \binom{N}{i} (\frac{\rho s_r}{S_x})^i (1 - \frac{\rho s_r}{S_x})^{N-i}, \ k \geqslant 1. \tag{2.12}$$

Roughly, we use $\check{\alpha}_k$ (or $\check{\alpha}_k(\rho)$) as the lower bound of the point $k$-coverage probability.

In Fig. 2.4, we plot the bounds of $k$-coverage probability ($k = 1, 2, 3$) as the number of nodes given. The area of interest is a $10 \times 10$ square and the sensing radius is 0.1. Hence $S = 100$ and $S_x = 104.04$.

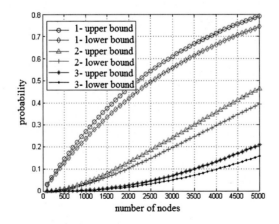

**FIGURE 2.4**

Bounds of 1, 2, 3-coverage probability ($L = 10, r = 0.1$).

Inversely, the formulas of probability bounds stipulate the range of the number of nodes required by any given coverage probability.

## 2.3 Point coverage in heterogeneous WSNs

In real applications, a WSN may consist of various types of senor nodes. In this section, for a heterogeneous sensor network, we analyze the point $k$-coverage probability and discuss performance optimization.

### 2.3.1 Point coverage probability

Consider a heterogeneous WSN which has $J$ types of sensor nodes. For type $A_j$, the sensing radius is $r_j$ and there are $n_j$ nodes in the network. We denote $r = (r_1, r_2, \cdots, r_J)$ and $n = (n_1, n_2, \cdots, n_J)$. The total number of nodes is $N$, namely $N = \sum_{i=1}^{J} n_i$.

With a boolean sensing model, the sensing range for type $A_j$ is $s_j = \pi r_j^2$. Moreover, the working probability for a type $A_j$ sensor is $\rho_j$. Denote $\rho = (\rho_1, \rho_2, \cdots, \rho_J)$.

Denote the probability that any point in the area of interest is covered by $m_j$ ($m_j \leq n_j$) nodes of type $A_j$ as $\theta(\rho_j, m_j, S)$. Based on the same analysis as in the previous section, $\theta(\rho_j, m_j, S)$ is

$$\theta(\rho_j, m_j, S) = \binom{n_j}{m_j} \left( \frac{\rho_j s_j}{S} \right)^{m_j} \left( 1 - \frac{\rho_j s_j}{S} \right)^{n_j - m_j}. \tag{2.13}$$

Moreover, denote the probability that any point in the area of interest is $k$-covered as $\Theta_k(\rho, S)$. Then we have

$$\Theta_k(\rho, S) = 1 - \sum_{m=0}^{k} \left\{ \sum_{\sum_{j=1}^{J} m_j = m, \, m_j \leq n_j} \prod_{j=1}^{J} \theta(\rho_j, m_j, S) \right\}, \tag{2.14}$$

where $\sum_{\sum_{j=1}^{J} m_j = m, \, m_j \leq n_j} \prod_{j=1}^{J} \theta(\rho_j, m_j, S)$ is the probability that any point is exactly covered by $m$ working nodes. The summation includes all the possible scenarios of node combinations. $\Theta_k(\rho, S)$ is also an upper bound of the point $k$-coverage probability if we are considering the boundary effect.

Denote $S_x = (S_x^{(1)}, S_x^{(2)}, \cdots, S_x^{(M)})$, where $S_x^{(j)} = (L + 2r_j)^2$. With Eq. (2.12), we get the lower bound of the probability for $k$-coverage in the heterogeneous WSN,

$$\check{\Theta}_k(\rho, S_x) = 1 - \sum_{m=0}^{k} \left\{ \sum_{\sum_{j=1}^{J} m_j = m, \, m_j \leq n_j} \prod_{j=1}^{J} \theta(\rho_j, m_j, S_x^{(j)}) \right\}. \tag{2.15}$$

For the same reason as for a homogeneous WSN, Eq. (2.14) also yields an upper bound of the $k$-coverage probability for a heterogeneous case.

**Example 2.3.** Suppose a WSN consists of type A and type B sensor nodes. Their sensing radii are $r_A = 15$ and $r_B = 10$, respectively. The total number of nodes is 5000. The side length of the area of interest is $L = 1000$. With Eq. (2.14) and Eq. (2.15), we get the upper and lower bounds of 2, 3-coverage probabilities when $n_A = 0, 100, \cdots, 5000$ and $n_B = 5000 - n_A$, as shown in Fig. 2.5.

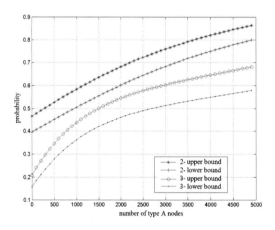

**FIGURE 2.5**

Bounds of 2, 3-coverage probability in a heterogeneous WSN.

## 2.3.2 Optimization

In a heterogeneous WSN, both the cost on sensor nodes and the network performance are subject to the combination of sensor nodes. Optimization aims to improve the coverage performance and/or reduce the cost on sensor nodes.

Suppose the type $A_j$ sensors cost $c_j$ $(j = 1, 2, \cdots, M)$ per node. Denote $c = (c_1, c_2, \cdots, c_M)$. The total cost on nodes, $B(c, n)$, is

$$B(c, n) = \sum_{j=1}^{M} c_j n_j. \tag{2.16}$$

Suppose the budget on the sensor nodes is limited by $\bar{B}$. To get the best performance with the budget, the optimization problem is formulated as follows:

$$\max \Theta_k(\rho, S)$$
$$s.t. \quad \begin{cases} B(c, n) \leq \bar{B}, \\ n_j \in [\underline{N_j}, \overline{N_j}], \ j = 1, 2, \cdots, M, \end{cases} \tag{2.17}$$

where $[\underline{N_j}, \overline{N_j}]$ is the predefined range for $n_j$.

Similarly, to optimize the lower bound of the point $k$-coverage probability, $\check{\Theta}_k(\rho, S_x)$, the formulation is

$$\max \check{\Theta}_k(\rho, S_x)$$
$$s.t. \quad \begin{cases} B(c, n) \leq \bar{B}, \\ n_j \in [\underline{N_j}, \overline{N_j}], \ j = 1, 2, \cdots, M. \end{cases} \quad (2.18)$$

**Example 2.4.** Type A and type B sensor nodes are to be deployed in a $1000 \times 1000$ square meters. Their price and sensing radius are shown in Table 2.1. The budget on nodes is \$50000. To maximize the point 3-coverage probability, which is the optimal combination?

**Table 2.1** Parameters of type A and type B sensor nodes.

| Type | $c$ | $r$ | $\rho$ |
|------|------|------|------|
| A | \$ 25 | 30 m | 1.0 |
| B | \$ 14 | 20 m | 1.0 |

Suppose $n_A$ and $n_B$ represent the numbers of type A and type B sensor nodes, respectively. $n_A \in [0, 50000/25]$ and $n_B \in [0, 50000/14]$. For each combination of $(n_A, n_B)$ subject to the constraint $25n_A + 14n_B \leq 50000$, Eq. (2.14) and Eq. (2.15) yield the upper and lower bounds of the point 3-coverage probability. All results are plotted in Fig. 2.6 where the X-axis is $n_A$ ($n_B = \lfloor(50000 - 25n_A)/14\rfloor$). When $n_A = 1400$, the upper bound of the point 3-coverage probability reaches the peak, the value is 0.9253. The upper bound and the lower bound may not be maximized with the same combination. In fact, the lower bound is maximized when $n_A = 1250$. Roughly speaking, for the heterogeneous WSN, any combination with $1250 \leq n_A \leq 1400$ and $n_B = \lfloor(50000 - 25n_A)/14\rfloor$ is a near-optimal solution to Problem (2.17) and its corresponding performance will be good enough.

On the other hand, if the requirement on point $k$-coverage probability is predefined, say $P^*$, to reduce the cost as much as possible, we should solve the following problem:

$$\min B(c, n)$$
$$s.t. \quad \begin{cases} \Theta_k(\rho, S) \geq P^*, \\ n_j \in [\underline{N_j}, \overline{N_j}], \ j = 1, 2, \cdots, M, \end{cases} \quad (2.19)$$

where $\Theta_k(\rho, S)$ is the upper bound of the point $k$-coverage probability. Or we may replace it by the lower bound, $\check{\Theta}_k(\rho, S_x)$,

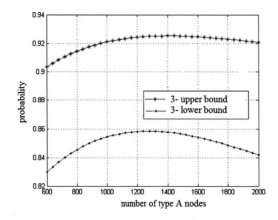

**FIGURE 2.6**

Bounds of point 3-coverage probability.

$$\begin{aligned} &\min B(c, n) \\ &s.t. \quad \begin{cases} \check{\Theta}_k(\rho, S_x) \geq P^*, \\ n_j \in [\underline{N_j}, \overline{N_j}], \ j = 1, 2, \cdots, M. \end{cases} \end{aligned} \tag{2.20}$$

According to the analysis in Section 2.2.3, the solution to Problem (2.19) may fail to reach $P^*$ at the boundary region; however, the solution to Problem (2.20) may be a little conservative.

## 2.4 Simulation experiments

First of all, the following sequence of commands with MATLAB® is used to randomly and uniformly distribute $N$ sensor nodes, say $O_1, O_2, \cdots, O_N$, in the area of interest:

```
for i = 1 : 1 : N
a = rand;
b = rand;
x(i) = a * L;
y(i) = b * L;
end
```

where the command "rand" returns a random number in [0, 1]. When the lower-left corner of the area of interest is assumed to be the origin, $(x(i), y(i))$ is the coordinate of node $O_i$.

Set $L = 1000$. The sensing radius will be specified case by case. To calculate the exact coverage rate, we divide the $1000 \times 1000$ square into $500 \times 500$ grids. The side length of each grid is 2. If the center of a grid is $k$-covered (i.e. the center falls in the sensing ranges of $k$ working nodes), we think this grid is $k$-covered. Then we divide the number of grids that are $k$-covered by $250,000$ to estimate the coverage probability.

### 2.4.1 Homogeneous WSNs

When the boundary effect is being ignored, Eq. (2.1) reveals the relationship between point coverage probability and the number of working nodes in a WSN. The bounds and the experimental values of point 2-coverage probability are shown in Fig. 2.7A and Fig. 2.7B when the sensing radii are $r = 20$ and $r = 100$, respectively. In simulations, for each given $N$, we have 100 independent runs and calculate the average probability as the experimental value. In Fig. 2.7A, the experimental values show little difference from the theoretical ones, while, in Fig. 2.7B, the experimental values are obviously smaller than the theoretical ones. The boundary effect is trivial when $r$ is much smaller than $L$.

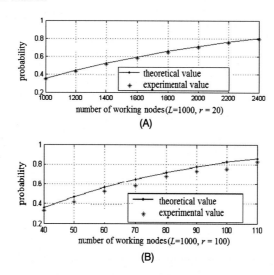

**FIGURE 2.7**

The boundary effect.

When the sensing radius $r = 50$, $N$ increases from 400 to 790 by 10 per step. For each fixed $N$, we replicate the same experiment for 100 times and get the point 1-coverage probability and 2-coverage probability on average. In Fig. 2.8, we plot the experimental results and compare them with the theoretical bounds with Eq. (2.9) and Eq. (2.11). Fig. 2.8 shows that, as $N$ is getting larger, the experimental values are getting closer to the theoretical bounds and the boundary effect gradually diminishes.

Fig. 2.9 shows the upper and lower bounds of the number of working nodes when the sensing radius $r = 50$ and the point 2-coverage and 3-coverage probabilities increase from 0.8 to 0.99 by 0.05 per step. When the point coverage probability is smaller than 0.94, it behaves approximately linearly with the number of working

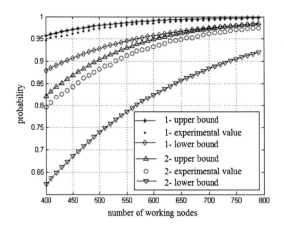

**FIGURE 2.8**

Experimental values and bounds of point 1, 2-coverage probability.

nodes. Therefore the point coverage provides a good tradeoff between the performance and the cost. When the point coverage probability is greater than 0.94, the number of working nodes increases rapidly.

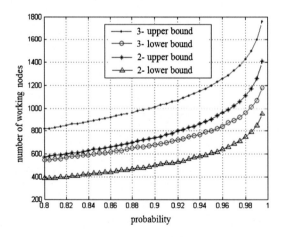

**FIGURE 2.9**

Bounds of the number of working nodes.

Fig. 2.8 shows that the experimental value is approximately the average of the upper and lower bounds of the point coverage probability. Therefore we may guess that the average of the upper and lower bounds of the number of working nodes may let the network achieve the desired coverage probability.

## 2.4.2 Heterogeneous WSNs

By simulation, we study a heterogeneous WSN which consists of two types of sensor nodes. The sensing radii of type A and type B are $r_A = 50$ and $r_B = 100$, respectively.

Let the number of type A nodes be 50 and the number of type B nodes increases from 10 to 150 by 10 per step. Fig. 2.10 shows the bounds and the experimental values of the point 1-coverage probability.

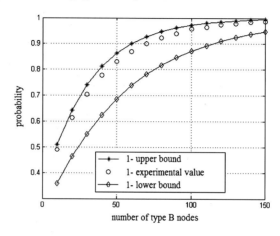

**FIGURE 2.10**

Experimental values and bounds of point 1-coverage probability.

Assume the price of type A sensor node is $c_A = \$20$ and the price of type B is $c_B = \$86$. The budget for the sensor nodes is \$5000. By solving Problem (2.18), we obtain the combinations which maximize the point 1-coverage probability. The top five $(n_A, n_B)$ combinations are presented in Table 2.2. The experimental value and the upper bound of point 1-coverage probability corresponding to each combination are also listed in Table 2.2. The experimental value is averaged over 100 independent runs.

**Table 2.2** Top five combinations when budget= \$5000.

| Combination | $(n_A, n_B)$ | Cost | Experimental value | Upper bound |
|:---:|:---:|:---:|:---:|:---:|
| 1 | (237, 3) | 4998 | 0.842675 | 0.859634 |
| 2 | (232, 4) | 4984 | 0.840636 | 0.858577 |
| 3 | (228, 5) | 4990 | 0.841966 | 0.858631 |
| 4 | (215, 8) | 4988 | 0.842712 | 0.857674 |
| 5 | (207, 10) | 5000 | 0.838651 | 0.857782 |

On the other hand, if the point 1-coverage probability is desired to be greater than or equal to 0.90, to minimize the cost on nodes, by solving Problem (2.20), we get the combinations which minimize the cost on sensor nodes. The top six $(n_A, n_B)$ combinations are presented in Table 2.3. The experimental value and the upper bound of the point 1-coverage probability corresponding to each combination are listed in Table 2.3. The experimental value is also averaged over 100 independent runs.

**Table 2.3** Top six combinations when point 1-coverage probability $\geq 0.90$.

| Combination | $(n_A, n_B)$ | Cost | Experimental value | Upper bound |
|---|---|---|---|---|
| 1 | (303, 5) | 6490 | 0.908200 | 0.921719 |
| 2 | (291, 8) | 6508 | 0.907263 | 0.921808 |
| 3 | (284, 10) | 6540 | 0.906806 | 0.922481 |
| 4 | (272, 13) | 6558 | 0.908204 | 0.922569 |
| 5 | (264, 15) | 6570 | 0.908970 | 0.922628 |
| 6 | (257, 17) | 6602 | 0.909874 | 0.923294 |

## 2.5 Summary

In this chapter, the point $k$-coverage probability in randomly deployed WSNs is analyzed. The boundary effect on the coverage performance is also studied. For heterogeneous WSNs, two optimization problems are formulated. One is to maximize the coverage probability with a given budget on sensor nodes; the other is to minimize the cost on the nodes while meeting the requirement on coverage. Simulation results show that the theoretical analysis of coverage performance is helpful for design and planning of WSNs.

## References

[1] H. Bai, X. Chen, Y.C. Ho, X. Guan, Information coverage configuration with energy preservation in large scale wireless sensor networks, in: Proceedings of IEEE International Conference on Computer and Information Technology, CIT'06, Seoul, Korea, September 20–22, 2006.
[2] X. Bai, S. Li, J. Xu, Mobile sensor deployment optimization for $k$-coverage in wireless sensor networks with a limited mobility model, IETE Technical Review 27 (2) (2010) 124.
[3] A. Chen, S. Kumar, T.H. Lai, Designing localized algorithms for barrier coverage, in: Proceedings of the 13th Annual ACM International Conference on Mobile Computing and Networking, 2007, pp. 63–74.
[4] X. Chen, Y.C. Ho, H.X. Bai, Complete coverage and point coverage in randomly distributed sensor networks, Automatica 45 (6) (2009) 1549–1553.
[5] T.L. Chin, P. Ramanathan, K.K. Saluja, K.C. Wang, Exposure for collaborative detection using mobile sensor networks, in: Proceedings of the 2nd IEEE International Conference on Mobile Ad-Hoc and Sensor Systems, MASS'05, Washington, DC, USA, November 7–10, 2005.
[6] T. Clouqueur, V. Phipatanasuphorn, P. Ramanathan, K.K. Saluja, Sensor deployment strategy for detection of targets traversing a region, ACM Mobile Networks and Applications 8 (2003) 453–461.
[7] P. Hall, Introduction to the Theory of Coverage Processes, John Wiley & Sons, 1988.
[8] N. Heo, P.K. Varshney, Energy-efficient deployment of intelligent mobile sensor networks, IEEE Transactions on Systems, Man and Cybernetics, Part A: Systems and Humans 35 (1) (2005) 78–92.

[9] S. Kumar, T.H. Lai, A. Arora, Barrier coverage with wireless sensors, in: Proc. ACM MobiCom, 2005.

[10] S. Kumar, T.H. Lai, J. Balogh, On $k$-coverage in a mostly sleeping sensor network, in: Proceedings of ACM Mobicom, 2004, pp. 144–158.

[11] A. Kwok, S. Martinez, Deployment algorithms for a power-constrained mobile sensor network, International Journal of Robust and Nonlinear Control 20 (7) (2010) 745–763.

[12] H. Ma, Y. Liu, On coverage problems of directional sensor networks, Mobile Ad-Hoc and Sensor Networks 3794 (2005) 721–731.

[13] S. Meguerdichian, F. Koushanfar, G. Qu, M. Potkonjak, Exposure in wireless ad hoc sensor networks, in: Proceedings of the 7th Annual ACM International Conference on Mobile Computing and Networking, MobiCom'01, Rome, Italy, July 16–21, 2001, pp. 139–150.

[14] W. Mo, D. Qiao, Z. Wang, Mostly-sleeping wireless sensor networks: connectivity, $k$-coverage, and $\alpha$-lifetime, in: Proceedings of the Allerton Conference, UIUC, 2005.

[15] V. Phipatanasuphorn, P. Ramanathan, Vulnerability of sensor networks to unauthorized traversal and monitoring, IEEE Transactions on Computers 53 (3) (2004) 364–369.

[16] M.R. Senouci, A. Mellouk, Deploying Wireless Sensor Networks: Theory and Practice, ISTE Press - Elsevier, 2016.

[17] G. Veltri, Q. Huang, G. Qu, M. Potkonjak, Minimal and maximal exposure path algorithms for wireless embedded sensor networks, in: Proceedings of the 1st International Conference on Embedded Networked Sensor Systems, SenSys'03, Los Angeles, California, USA, November 5–7, 2003, pp. 40–50.

[18] P.J. Wan, C.W. Yi, Coverage by randomly deployed wireless sensor networks, in: IEEE International Symposium on Network Computing and Applications, 2005, pp. 275–278.

[19] B. Wang, K.C. Chua, V. Srinivasan, W. Wang, Sensor Density for Complete Information Coverage in Wireless Sensor Networks, Lecture Notes in Computer Science, vol. 3868, Springer, Berlin, 2006, pp. 69–82.

[20] B. Wang, W. Wang, V. Srinivasan, K.C. Chua, Information coverage for wireless sensor networks, IEEE Communication Letters 9 (11) (2005) 967–969.

[21] C.H. Wu, Y.C. Chung, A polygon model for wireless sensor network deployment with directional sensing areas, Sensor 9 (2009) 9998–10022.

[22] H. Zhang, J.C. Hou, On the upper bound of $\alpha$-lifetime for large sensor networks, The ACM Transactions on Sensor Networks 1 (2) (2005) 272–300.

CHAPTER

# Percentage coverage schemes

3

## CONTENTS

Complete coverage can guarantee that the area of interest is fully covered without any vacancy. However, in many applications, complete coverage is unnecessary and a small part of sensing loss is acceptable. Moreover, relaxing the requirement from complete coverage to a percentage coverage can save much energy.

In this chapter, a location-based Percentage Coverage Configuration Protocol (PCCP) is developed to guarantee a certain percentage of the area of interest covered while some sensor nodes are scheduled sleeping (or OFF). The concept of the sensor's occupation area is introduced to implement the idea of divide-and-conquer. Analysis and simulation demonstrate the performance of PCCP [3]. Moreover, coverage configuration in the absence of sensor locations is also studied. A distributed, localized and location-free node-scheduling algorithm, the Standing Guard Protocol (SGP), is proposed to deal with coverage and connectivity within a unified framework. Simulation results show that SGP can achieve the predefined coverage percentage and maintain a good balance between coverage performance and energy-consumption of the WSN [4].

**Randomly Deployed Wireless Sensor Networks. https://doi.org/10.1016/B978-0-12-819624-3.00008-2**
Copyright © 2020 Tsinghua University Press. Published by Elsevier Inc. All rights reserved.

## 3.1 Location-based percentage coverage

Much work on coverage configuration focuses on coverage preserving which means, even with some sensor nodes scheduled OFF, there should be no loss of coverage within the area of interest. Tian and Georganas [19] present a coverage-preserving node-scheduling scheme to maintain complete coverage. Jiang and Dou [12] propose a decentralized and localized coverage-preserving density control algorithm to prolong network lifetime. Wang et al. [21] present a coverage configuration protocol which can provide a different degree of coverage and analyze the relationship between coverage and connectivity. Ye et al. [22] consider the sensor node failure rate and propose a node-scheduling algorithm to ensure that every point is being monitored by working nodes. Hsin and Liu [11] develop a random and coordinated sleep algorithm to implement network coverage using low duty-cycled sensor nodes. However, all of the above work cannot meet a flexible percentage coverage requirement.

The analysis in Chapter 2 shows that relaxing the requirement from complete coverage to partial coverage can prolong the network lifetime. Zhang and Hou [23] have derived that the upper bound of the lifetime can increase by 15% for 99%-coverage and by 20% for 95%-coverage. Fig. 3.1 explains the reason intuitively, where the square is the area of interest and three discs represent the sensing region of three sensor nodes (marked with '•'). To completely cover the square, all of the three sensor nodes have to work. If a small vacancy in coverage is acceptable, one or two of them may be OFF.

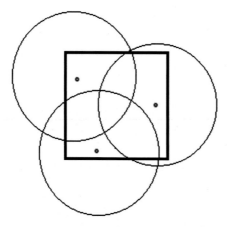

**FIGURE 3.1**

Complete coverage and partial coverage.

There is less work on percentage coverage and much less on location-based percentage coverage. Because percentage coverage on the area of interest is a global concept, distributed and localized coverage configuration can hardly achieve the exact percentage. To overcome this difficulty, we introduce the concept of the occupation area to implement the idea of divide-and-conquer. Moreover, we let sensor nodes cluster and let each cluster take charge of an occupation area. In this way, if each cluster lets the covered area of its occupation area achieve the desired percentage, then the total covered area in the whole area of interest also reaches the percentage. With this idea, we propose a distributed and localized PCCP.

### 3.1.1 Occupation area

The important concept used in PCCP is Voronoi diagram. Suppose there are $N$ sensor nodes in a 2D plane, if we partition the plane into $N$ convex polygons such that each polygon contains exactly one node and every point in a given polygon is closer to the node in this polygon than to any other node, then we get a Voronoi diagram [1], as shown in Fig. 3.2 for instance.

In a Voronoi diagram, each polygon is called a Voronoi cell. Particularly, we call a sensor node's Voronoi cell its occupation area. If two occupation areas share a common edge, the owners of the two occupation areas are called Voronoi neighbors.

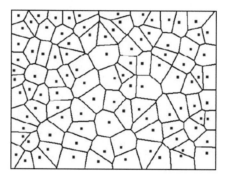

**FIGURE 3.2**

Illustration of Voronoi diagram.

Suppose there are $N$ sensor nodes, say $O_1, O_2, \cdots, O_N$, in the network. Denote by $OA(O_i)$ the occupation area of node $O_i$ ($1 \le i \le N$). Then we have

$$OA(O_i) = \{a | d(O_i, a) \le d(O_j, a), \forall j \ne i, 1 \le j \le N\}, \tag{3.1}$$

where $a$ is any point in the area of interest. Denote the position of $a$ as $(x_a, y_a)$. For simplicity and without loss of generality, suppose the area of interest is a unit square. Then $x_a$ and $y_a$ satisfy $x_a \ge 0, y_a \ge 0, x_a \le 1, y_a \le 1$.

Denote the position of node $O_i (1 \le i \le N)$ as $(\gamma_i^x, \gamma_i^y)$. Then $d(O_i, a) \le d(O_j, a)$ may be written as

$$\sqrt{(x_a - \gamma_i^x)^2 + (y_a - \gamma_i^y)^2} \le \sqrt{(x_a - \gamma_j^x)^2 + (y_a - \gamma_j^y)^2},$$

that is,

$$2(\gamma_j^x - \gamma_i^x)x_a + 2(\gamma_j^y - \gamma_i^y)y_a + [(\gamma_i^x)^2 + (\gamma_i^y)^2 - (\gamma_j^x)^2 - (\gamma_j^y)^2] \le 0.$$

Therefore OA($O_i$) may be written as

$$
\text{OA}(O_i) =
\left\{
a = (x_a, y_a) \middle|
\begin{array}{c}
2(\gamma_j^x - \gamma_i^x)x_a + 2(\gamma_j^y - \gamma_i^y)y_a \\
+[(\gamma_i^x)^2 + (\gamma_i^y)^2 - (\gamma_j^x)^2 - (\gamma_j^y)^2] \leq 0, \\
0 \leq x_a \leq 1, 0 \leq y_a \leq 1, \forall j \neq i, 1 \leq j \leq N.
\end{array}
\right\}
\tag{3.2}
$$

Eq. (3.2) is the exact definition of node $O_i$'s occupation area. To compute a sensor node's occupation area with Eq. (3.2), one needs to know the locations of the remaining nodes. This is almost impossible for a large-scale sensor network. Fortunately, in fact, when a sensor node is computing its occupation area, the node only needs to know its neighbors' locations (see Fig. 3.2).

In a WSN, if two sensor nodes can communicate with each other, then they are neighbors. If we denote by $R$ the communication radius, then a node's communication range is a disc with radius $R$ and centered at itself.

In general, we impose the following assumption on the communication radius.

**Assumption 3.1.** *The area of interest is a unit square.*

**Assumption 3.2.** *Sensor nodes are randomly and uniformly distributed in the area of interest.*

**Assumption 3.3.** *A Boolean sensing model is adopted.*

**Assumption 3.4.** *The communication radius, R, is larger than two times of the sensing radius, r, i.e. $R \geq 2r$.*

Suppose node $O_i$ has $M$ neighbors, which are denoted as $O_{i1}, O_{i2}, \cdots, O_{iM}$. Their positions are $(\gamma_{im}^x, \gamma_{im}^y)$, $1 \leq m \leq M$, respectively. Node $O_i$ may compute its occupation area merely based on the positions of its neighbors instead of all other sensor nodes in the network. Such an occupation area is called a neighbor-based occupation area; it is denoted as NOA($O_i$). Then we have

$$
\text{NOA}(O_i) =
\left\{
a = (x_a, y_a) \middle|
\begin{array}{c}
2(\gamma_{im}^x - \gamma_i^x)x_a + 2(\gamma_{im}^y - \gamma_i^y)y_a \\
+[(\gamma_i^x)^2 + (\gamma_i^y)^2 - (\gamma_{im}^x)^2 - (\gamma_{im}^y)^2] \leq 0, \\
0 \leq x_a \leq 1, 0 \leq y_a \leq 1, 1 \leq m \leq M.
\end{array}
\right\}
\tag{3.3}
$$

Compared with Eq. (3.2), Eq. (3.3) has less inequalities for NOA($O_i$). Therefore NOA($O_i$) $\supseteq$ OA($O_i$). The value of the neighbor-based occupation area is greater than or equal to the value of the occupation area. Denote by SA($O_i$) the sensing area of node $O_i$. We have the following proposition.

**Proposition 3.1.** *For any sensor in the area of interest, $O_i$, and any point $a \in$ SA($O_i$), if $a \in$ OA($O_i$), then $a \in$ NOA($O_i$), and vice versa.*

*Proof.* Since $\text{NOA}(O_i) \supseteq \text{OA}(O_i)$, if $a \in \text{OA}(O_i)$, then we have $a \in \text{NOA}(O_i)$.

Now we prove $a \in \text{OA}(O_i)$ if $a \in \text{NOA}(O_i)$ by contradiction. Since $a \in \text{SA}(O_i)$, we have $d(O_i, a) \le r$. If $a \in \text{NOA}(O_i)$ but $a \notin \text{OA}(O_i)$, there must exist another node $O_j$ such that $a \in \text{OA}(O_j)$. Therefore $d(O_j, a) < d(O_i, a) \le r$, then we have $d(O_i, O_j) \le d(O_i, a) + d(O_j, a) < 2r$. With Assumption 3.4, node $O_j$ is a neighbor of node $O_i$, and then $a \in \text{NOA}(O_j)$. This is contradiction to $a \in \text{NOA}(O_i)$.   □

Fig. 3.3 helps us to understand the fact revealed by Proposition 3.1. In Fig. 3.3, each sensor node is marked with '•' and the shadowed area represents the sensing area of $O_1$, $r$ and $R$ are the sensing radius and the communication radius, respectively. In Fig. 3.3A, the Voronoi cell, the quadrangle $ABDC$, is the occupation area of node $O_1$. In Fig. 3.3B, since the distance between $O_2$ and $O_1$ is larger than the communication radius, $O_2$ is not a neighbor of $O_1$ but $O_3$, $O_4$ and $O_5$ are neighbors. Hence the neighbor-based occupation area of $O_1$ is the triangle $ABE$. According to Assumption 3.4, the part of quadrangle $ABDC$ in shadow is the same as the part of triangle $ABE$ in shadow. Proposition 3.1 tells us that even though a sensor node's occupation area may differ from its neighbor-based occupation area, the parts in the node's sensing area are exactly the same.

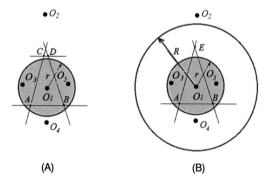

(A)                 (B)

**FIGURE 3.3**

Occupation area and neighbor-based occupation area.

The following proposition shows the condition for which a sensor node's occupation area is the same as its neighbor-based occupation area.

**Proposition 3.2.** *For a WSN, if the area of interest is completely 1-covered, then each sensor node's occupation area is the same as its neighbor-based occupation area.*

*Proof.* Since the area of interest is completely 1-covered, for any point $a$, there exists at least one sensor node, say $O_i(1 \le i \le N)$, such that the distance between $a$ and $O_i$, $d(O_i, a)$, satisfies $d(O_i, a) < r$. Hence each node's occupation area must fall in its sensing area; otherwise, there would exist at least one point not covered by any sensor node. Moreover, with Proposition 3.1, we complete the proof.   □

In a WSN which consists of a large number of sensor nodes, the area of interest is likely to be fully covered. In this case, each sensor node's neighbor-based occupation area is exactly the same as its occupation area.

### 3.1.2 Percentage coverage configuration protocol

Besides Assumptions 3.1-3.4, we impose some more assumptions as follows.

**Assumption 3.5.** *All sensor nodes are time synchronized.*

**Assumption 3.6.** *Each sensor node knows its own position as well as the boundary of the area of interest.*

**Assumption 3.7.** *Each sensor node has an unique ID.*

Both time synchronization and localization in WSNs are important issues; the reader may refer to [8,13,14] for time synchronization methods and refer to [6,15,17] for sensor node localization. Each node may use its position as its ID. Or each node simply independently generates a random number as its ID.

The basic idea of the Percentage Coverage Configuration Protocol (PCCP) is divide-and-conquer. All occupation areas constitute the area of interest without overlap. The area of interest is divided into a number of subregions and each subregion is a collection of occupation areas. Therefore the coverage configuration could be carried out within each subregion.

PCCP also adopts the way of working in rounds (see Section 2.1.4). Each round begins with a configuration phase for scheduling.

The node-scheduling procedure is implemented in two steps. The first one is information exchange and neighbor-based occupation area computation. Neighboring sensor nodes exchange their IDs and locations by communication (to avoid collision, each node should generate a random back-off time for broadcasting). Then each node computes its own neighbor-based occupation area and carries out a coverage configuration. The configuration procedure in detail is as follows.

(1) Competing to be a head.

Node $O_i (i = 1, 2, \cdots, N)$ generates a random back-off time $\delta_i$. As $\delta_i$ expires, node $O_i$ broadcasts a configuration starting message to declare itself as a head and then begins its work. If node $O_i$ has heard such a message from another node, say node $O_{h_i}$, before the time $\delta_i$ expires, then node $O_i$ gives up competing immediately and becomes a member of the head $O_{h_i}$ and then begins its work.

(2) Work for a member.

A member computes the area of its neighbor-based occupation area (denoted by $S_{NOA}$), the area of the part of its neighbor-based occupation area which is covered by its sensing area (denoted by $S_s$), and the area of the part of its neighbor-based occupation area covered by its head (denoted by $S_c$) and sends its $S_{NOA}$, $S_s$, $S_c$ to its head. After that, this sensor node will listen to the channel until it receives an off-duty eligible neighbor list from its head. If it is included in the list, it will turn OFF; otherwise, it will work in the round.

Usually, there are multiple members sharing a common head. Again, to avoid collision, each member should send its message with a random back-off time. The back-off time should be bounded by a given number, say $\bar{\delta}$, so that each head is sure that all of its members have finished information exchange. In addition, the computation time required by each member should also be upper bounded by a number, say $\bar{\tau}$, which could be known in advance.

(3) Work for a head.

A head also calculates the area of its neighbor-based occupation area (denoted by $\hat{S}_{\mathrm{NOA}}$), the area of the part of its neighbor-based occupation area which is covered by its sensing area (denoted by $\hat{S}_s$). Before the time $(\bar{\delta} + \bar{\tau})$ expires, the head listens and collects the information of $S_{\mathrm{OA}}$ and $S_c$ of its members. Then it executes the following percentage coverage configuration algorithm to decide the off-duty eligible neighbors. Finally, it broadcasts the list which includes all IDs of the off-duty eligible neighbors and stays working within the round.

**Algorithm 3.1** (Percentage coverage configuration).

Step 1: The head calculates the total area of its members' $S_{\mathrm{NOA}}$ and its own $\hat{S}_{\mathrm{NOA}}$,

$$\tilde{S} = \hat{S}_{\mathrm{NOA}} + \sum_{\text{all members}} S_{\mathrm{NOA}}.$$

Step 2: The head sorts its members in an ascending order by $S_c/S_{\mathrm{NOA}}$; the order is denoted as $O_1, O_2, \cdots, O_K$, where $K$ is the number of its members. The corresponding sequences of $S_c$ and $S_s$ are denoted as $S_c(1), S_c(2), \cdots, S_c(K)$ and $S_s(1), S_s(2), \cdots, S_s(K)$. The head computes the coverage percentage as follows:

$$\bar{P}_k = \frac{\pi r^2 - \sum_{i=1}^{k} S_c(i) + \sum_{i=1}^{k} S_s(i)}{\tilde{S}}, \quad k = 1, 2, \cdots, K, \tag{3.4}$$

where $\pi r^2$ is the sensing area of the head.

Step 3: The head takes $k^*$, which satisfies $\bar{P}_{k^*-1} < P^* \le \bar{P}_{k^*}$, where $P^*$ is the desired coverage percentage. Then members $O_1, O_2, \cdots, O_{k^*}$ will work within the round and the rest nodes $O_{k^*+1}, \cdots, O_K$ will be OFF. The head broadcasts an off-duty eligible neighbor list. In particular, if a member has $S_c/S_{\mathrm{NOA}} > P^*$, which means its head's sensing area covers a part of more than $P^*$ of its neighbor-based occupation area, then this member can be set OFF.

Note that for a member the part of its neighbor-based occupation area covered by its head (the area is $S_c$) must be in its own sensing area, since any point in this part is closer to the member. Moreover, the member and its head have the same sensing radii. Hence $S_c \le S_s$. Since a member's $S_s$ is smaller than its sensing area, the coverage percentage obtained by Eq. (3.4) is a lower bound of the coverage percentage. This coverage configuration protocol guarantees the desired percentage $P^*$.

### 3.1.3 Simulation and analysis

#### *Performance test*

We first simulate PCCP and evaluate its performance with coverage percentage and network lifetime. In our simulation, sensing radius $r = 0.1$; communication radius $R = 0.2$; the number of nodes $N = 500$; each node is able to work continuously for $L_0 = 1$ units of time.

To calculate the exact coverage percentage, we divide the area of interest (a unit square) into $500 \times 500$ grids. We estimate the coverage percentage with the number of grids whose centers are covered.

We randomly deploy 500 sensor nodes in the area of interest. For each node, independently generate two random numbers $x$ and $y$, both of them follow a uniform distribution on $(0, 1)$ and assign $(x, y)$ as the node's location. Set the desired coverage percentage $P^* = 0.50$. After the deployment, we simulate PCCP for 10 rounds and calculate the coverage percentage and the number of working nodes in each round. We replicate such a simulation 50 times. Then, over the $50 \times 10$ rounds, we get the average percentage, $\bar{P}$, and the average number of working nodes, $\bar{N}_w$. Roughly, we view $N/\bar{N}_w$ as an estimate of the network lifetime. Both $\bar{P}$ and $N/\bar{N}_w$ measure the performance of PCCP.

When $P^* = 0.50$, by simulation we get $\bar{P} \approx 0.77$ and $N/\bar{N}_w \approx 16$. $\bar{P} > P^*$ indicates that PCCP is a rather conservative protocol. Increasing $P^*$ from 0.50 to 0.96, for each value of $P^*$, we replicate the above simulation and get the corresponding $\bar{P}$ and $N/\bar{N}_w$ as shown in Fig. 3.4.

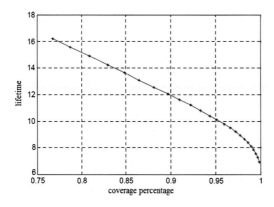

**FIGURE 3.4**

Network lifetime vs. coverage percentage.

Fig. 3.4 shows that, as the coverage percentage is close to 1 (i.e. complete coverage), the network lifetime drops fast. As coverage percentage drops from 0.99 to 0.95, the network lifetime increases by around 25%. Hence, given the number of nodes, partial coverage can efficiently prolong the network lifetime.

### Performance comparison: PCCP and CCP

Now we compare the performance of PCCP to the Coverage Configuration Protocol (CCP), which is presented in [21]. In our simulation, the sensing radius $r = 0.1$, the number of nodes $N = 200, 400, 600, 800, 1000$, and $P^* = 0.7, 0.8, 0.9$. $N$ is the parameter for CCP; the pair $(N, P^*)$ is the parameter for PCCP.

For each value of the parameter, we carry out 20 simulations. In each simulation, each node's location is randomly generated and then run 10 rounds. Over the $20 \times 10$ rounds, we get the average percentage, $\bar{P}$, and the average number of working sensor nodes, $\bar{N}_w$. $N / \bar{N}_w$ is viewed as an estimate of the network lifetime. Simulation results are shown in Figs. 3.5 and 3.6.

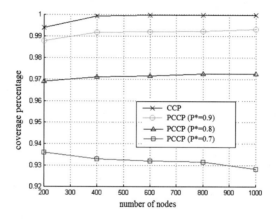

**FIGURE 3.5**

Coverage percentage.

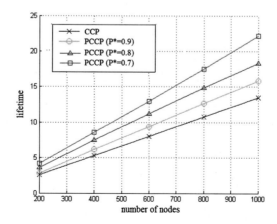

**FIGURE 3.6**

Network lifetime.

Compared to CCP, PCCP needs less working nodes and extends the network lifetime. For example, when $N = 1000$, the network lifetime for complete coverage is 13 units of time; but for a percentage coverage with $P^* = 0.7$, the lifetime increases by 69.2% to 22 units (see Fig. 3.6) and the coverage percentage is about 10% less (see Fig. 3.5).

## 3.2 Location-free percentage coverage

Much research has proposed various approaches for node localization (see e.g. [2,9, 10,15,17,18]). These approaches either need some special infrastructures (e.g. GPS) or massive communication, or both of them. Moreover, the localization accuracy may deteriorate in harsh physical environments.

Location-free coverage configuration protocols have a broad application background. Some location-free coverage configuration protocols have been developed to get complete coverage (see e.g. [22]). Tian and Georganas [20] study nearest-neighbor-based, neighbor-number-based and probability-based node-scheduling schemes. All of them work in rounds but do not guarantee complete coverage. A nearest-neighbor-based scheme lets a sensor node turn OFF if its nearest working neighbor is within a distance threshold. To avoid two neighbors turning OFF simultaneously, each node has to generate a random back-off time and broadcasts a message to inform other sensors that it will turn OFF when the back-off time expires. The neighbor-number-based scheme lets a sensor node turn OFF if the number of its neighbors is on a threshold. This scheme also uses the trick of a random back-off time. A probability-based algorithm lets a sensor node generate a uniform random number in $[0, 1)$; if the number is below a threshold, the node will turn OFF; otherwise it will keep working. Such a threshold is called an off-duty probability; it is equivalent to $1 - \rho$ in Section 2.2.1.

In this section, we present a location-free coverage configuration scheme, called a Standing Guard Protocol (SGP), based on the occupation area which is defined in Section 3.1.1. The basic idea is that working nodes will get in charge of the occupation areas of off-duty nodes. SGP also works in rounds. At the beginning of each round, every sensor node competes to be a 'guard' and the successful one will broadcast a message to declare its role. Other nodes listen and count the messages to decide whether to be OFF or keep working [4].

### 3.2.1 Occupation area

According to the definition of occupation area in Section 3.1.1, for a large-scale WSN, the area of a node's occupation area decreases as the number of nodes increases. For a node, say $O_1$, if the area of its occupation area is much smaller than its sensing area, denote the maximal distance between $O_1$ and any point in $OA(O_1)$ as $l$, then $l$ is smaller than its sensing radius $r$, i.e. $l < r$. Suppose a working node, say $O_2$, is $O_1$'s neighbor. $d(O_1, O_2)$ means the distance between $O_1$ and $O_2$.

When $l + d(O_1, O_2) < r$, OA$(O_1)$ will be completely covered by $O_2$'s sensing area, SA$(O_2)$, i.e. OA$(O_1) \subset$ SA$(O_2)$, as shown in Fig. 3.7. In this case, node $O_1$ turns OFF and its occupation area will be charged by node $O_2$. This is expressed by stating that working node $O_2$ stands for the OFF node $O_1$.

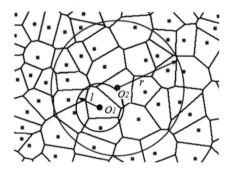

**FIGURE 3.7**

Node $O_1$'s occupation area covered by node $O_2$'s sensing area.

In this way, without the knowledge of sensor's locations, each sensor node can decide whether to switch OFF or stay working once it gets its own $l$ and $d$ (the distance to the nearest working node). Here, $l$ can be estimated according to node's density in a WSN, and $d$ can be estimated according to the predefined broadcasting radius, $R_{SG}$ (used exclusively for location-free coverage configuration). As working nodes have broadcast a standing guard message, any node which hears such a message is located within the distance of $R_{SG}$.

The basic idea of SGP is as follows. To obtain the coverage percentage $P^*$, let the working nodes broadcast standing guard message with radius $R_{SG} = r/\sqrt{P^*}$, nodes which hear the message will turn OFF. In Fig. 3.8, the shadowed part is a standing guard occupation area of working node $O_1$. For a large-scale WSN, the number of nodes, $N$, is usually very large. Hence $l$ is small. The area of the shadowed part is approximated to $\pi R_{SG}^2 = \pi r^2/P^*$. As the area of node $O_1$'s sensing area is $\pi r^2$, the coverage percentage for this part is around $P^*$. Moreover, as shown in Fig. 3.8, node $O_1$ and node $O_2$ are both working and the overlap of their standing guard occupation areas makes the coverage percentage greater than $P^*$.

### 3.2.2 Standing guard protocol

Based on Assumptions 3.1–3.3 and Assumptions 3.5 and 3.7, we impose one more assumption as follows.

**Assumption 3.8.** *Each sensor node has an adjustable communication radius.*

This assumption is reasonable (e.g. the MICA system [7]). We consider a WSN which consists of $N$ sensor nodes deployed randomly and uniformly in the area of interest, which is a unit area.

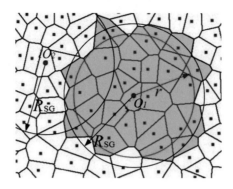

**FIGURE 3.8**

Illustration of standing guard area.

### Description of SGP

At the beginning of each round, all sensor nodes switch on and compete to be standing guard nodes. Each node executes the following procedure to implement coverage configuration.

Step 1: Initialization. Generating a random back-off time $\delta_o$ ($\delta_o$ is upper bounded by $\bar{\delta}$, which is much smaller than the period length).

Step 2: Once receiving a standing guard message within $\delta_o$, switching OFF immediately; otherwise, when $\delta_o$ expires, broadcasting standing guard message (including own ID) with communication radius $R_{SG}$ and remaining working in this round.

With these two steps, each node completes its 1-degree percentage coverage configuration. However, the values of the parameters $\bar{\delta}$ and $R_{SG}$ need to be determined. As all sensor nodes need to switch on for configuration, a larger $\bar{\delta}$ will cost more energy but a smaller $\bar{\delta}$ will more likely result in communication collisions. When choosing the value of $\bar{\delta}$, we need to consider the node density in the WSN as well as the tradeoff between energy saving and collision avoidance. Below, we will discuss how to choose the value of $R_{SG}$.

SGP is different from the Probing Environment and Adaptive Sleeping (PEAS) protocol presented in [22] or the nearest-neighbor-based scheme in [20] although they have some similar features. PEAS lets each node probe the working nodes in its neighborhood; SGP only allows working nodes to broadcast a standing guard message; the nearest-neighbor-based scheme lets all sensor nodes first exchange information by communication and then lets those nodes be OFF and they will broadcast a duty-free message. Obviously, the communication load in SGP is the least one and hence SGP is most energy-efficient.

### Analysis of SGP performance

Similar to the location-based percentage coverage in Section 3.1, we require that the coverage percentage is greater than or equal to $P^*$.

Bai et al. [4] provide the condition of SGP for complete $k$-coverage. Readers may refer to [4] for the performance of SGP when $k > 1$. In the following, we only discuss SGP when $k = 1$. We first analyze the number of working nodes.

**Proposition 3.3.** *After the configuration with SGP, the number of working nodes in WSN is around* $N_{SG} = 8/(3\sqrt{3}R_{SG}^2)$.

*Proof.* According to SGP, in the area of interest, the distance between any two working nodes is greater than or equal to $R_{SG}$. Now we divide the area of interest into a number of equivalent regular hexagons and the side length is $R_{SG}/2$. The area of each hexagon is $3\sqrt{3}R_{SG}^2/8$. As the distance between any two points in such a hexagon is shorter than $R_{SG}$, at most there is one working sensor in it. To cover the area of interest with such hexagon, the number of hexagons is about $N_{SG} = 1/(3\sqrt{3}R_{SG}^2/8) = 8/(3\sqrt{3}R_{SG}^2)$. □

The area of interest in reality may not be a unit square. If its area is $S$, then the number of working nodes in the WSN is around $8S/(3\sqrt{3}R_{SG}^2)$.

Proposition 3.3 reveals that the number of working nodes scheduled by SGP hinges on $R_{SG}$ and is independent on $N$ when $N$ is large enough. Therefore SGP is good for scalability and applicable to large-scale WSNs. The remaining question is how to choose the standing guard radius $R_{SG}$.

Since there are $N$ sensor nodes randomly and uniformly distributed in the area of interest, after coverage configuration, we assume that the working nodes are also uniformly distributed in the area of interest. Suppose the network needs $M_0$ working nodes to make the point 1-coverage probability reach $P^*$. With Eq. (2.1), we have

$$\alpha_1 = 1 - (1 - \pi r^2)^{M_0} \geq P^*.$$

Hence, $M_0$ should satisfy

$$M_0 \geq \frac{\ln(1 - P^*)}{\ln(1 - \pi r^2)}.$$

With Proposition 3.3, if we apply SGP with communication radius $R_{SG}$ for configuration, the number of working nodes in the network is about $8/(3\sqrt{3}R_{SG}^2)$. Therefore, we let

$$8/(3\sqrt{3}R_{SG}^2) \geq \frac{\ln(1 - P^*)}{\ln(1 - \pi r^2)},$$

that is,

$$R_{SG} \leq \sqrt{\frac{8\ln(1 - \pi r^2)}{3\sqrt{3}\ln(1 - P^*)}}.$$

As $r$ is much smaller than 1, $\pi r^2$ is also smaller than 1, so we have $\ln(1 - \pi r^2) \approx -\pi r^2$; moreover, $8\pi/(3\sqrt{3}) \approx 5$, and we obtain

$$R_{SG} \leq \frac{\sqrt{5}r}{\sqrt{|\ln(1 - P^*)|}}. \tag{3.5}$$

Based on the above analysis, we have the following proposition to determine the value of $R_{SG}$.

**Proposition 3.4.** *Suppose the number of nodes, N, is big enough. If SGP is applied for configuration and $R_{SG} \leq \dfrac{\sqrt{5}r}{\sqrt{|\ln(1 - P^*)|}}$, then the point 1-coverage probability is greater than or equal to $P^*$.*

With this proposition, given either $R_{SG}$ or $P^*$, the other one can be obtained. For example, if $R_{SG} = 2r$, then $P^* = 71.3\%$; if the coverage percentage is required to be $P^* = 99.3\%$, then $R_{SG} \leq r$, which means the standing guard radius is about equal to the sensing radius.

Now we analyze the network lifetime. Suppose each node is able to work continuously for $L_0$ units of time and $R_{SG} = \dfrac{\sqrt{5}r}{\sqrt{|\ln(1 - P^*)|}}$. Then, with Proposition 3.3, the number of working nodes is

$$N_{SG} = \frac{8}{3\sqrt{3}R_{SG}^2} = \frac{8|\ln(1 - P^*)|}{15\sqrt{3}r^2}.$$

If ignoring the energy cost on the node's state switching and configuration, then we get an upper bound for the lifetime, which is

$$\hat{L}_{sn} = \left(\frac{N}{N_{SG}}\right)L_0 = \left(\frac{15\sqrt{3}r^2 N}{8|\ln(1 - P^*)|}\right)L_0.$$

The network lifetime $\hat{L}_{sn}$ is proportional to the number of nodes $N$. In other words, a larger $N$ results in a longer lifetime.

In general, if the area of the area of interest is $S$ and the number of nodes in the network is still $N$, the network lifetime becomes

$$\hat{L}_{sn}(S) = \left(\frac{15\sqrt{3}r^2 N}{8S|\ln(1 - P^*)|}\right)L_0.$$

Conversely, to make the network work for $L_{sn}^*$ units of time, the number of nodes in the network should be

$$\hat{N}(S) = \left(\frac{8S|\ln(1 - P^*)|}{15\sqrt{3}r^2 N}\right)\frac{L_{sn}^*}{L_0}.$$

### 3.2.3 Simulation and analysis
#### *Performance test*
In the simulation, the sensing radius $r = 0.1$; each sensor node is able to work continuously for $L_0 = 1$ units of time; the number of nodes $N = 500, 1000, 1500, 2000, 2500, 3000$.

To calculate the exact coverage percentage, we divide the area of interest (a unit square) into $500 \times 500$ grids. We use the number of grids whose centers are covered to estimate the coverage percentage. We test the performance when coverage percentage $P^* = 0.99$ and $P^* = 0.90$. With Proposition 3.4, we get the corresponding communication radii; they are $R_{SG} = 0.1005$ and $R_{SG} = 0.1054$, respectively. The radius for $P^* = 0.99$ is a little less than the one for $P^* = 0.90$. A smaller communication radius will let more nodes work to achieve a greater coverage percentage.

For each $N$, we randomly distribute $N$ sensor nodes in the area of interest and then simulate SGP with $P^* = 0.99$ and $P^* = 0.90$ separately for 10 rounds. In each round, we calculate the coverage percentage and the number of working nodes. We replicate such a simulation 50 times. At last, over the $50 \times 10$ rounds, we get the average percentage $\bar{P}$ and the average number of working sensors $\bar{N}_w$. Moreover, we use $N/\bar{N}_w$ as an estimate of the network lifetime. The simulation results are plotted in Fig. 3.9 and Fig. 3.10, respectively.

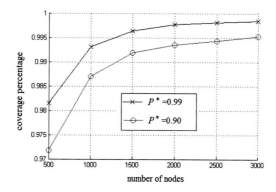

**FIGURE 3.9**

Coverage percentage.

Fig. 3.9 shows that, when $N = 500$, $P^* = 0.99$ is not satisfied, since the network has not enough sensor nodes. As $N$ increases, the resulting coverage percentage may exceed 0.99.

### Performance comparison: SGP and the three schemes in [20]

Now we compare SGP with the three location-free configuration schemes presented in [20], which are nearest-neighbor-based, neighbor-number-based, and probability-based schemes.

In our simulation, we set $P^* = 0.8$. In [20], the parameters corresponding to $P^* = 0.8$ are as follows. For the nearest-neighbor-based scheme, the distance threshold $D = 0.315, r = 0.0315$; for the neighbor-number-based scheme, the threshold for the number of neighbors $\Upsilon = 6$ when the communication radius $R = 0.2$; and for the probability-based algorithm, the off-duty probability can be obtained with $\Upsilon$.

Taking $N = 200, 400, 600, 800, 1000$, for each of the values, we have 100 independent runs. In the simulation, each node's location is randomly generated and then

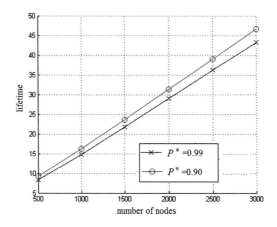

**FIGURE 3.10**

Network lifetime.

run 10 rounds. Finally, over the $100 \times 10$ rounds, we compute the average percentage $\bar{P}$ and the average number of working sensors $\bar{N}_w$. $N/\bar{N}_w$ is viewed as an estimate of the network lifetime. Simulation results are shown in Figs. 3.11 and 3.12.

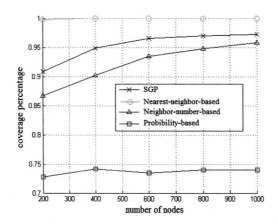

**FIGURE 3.11**

Coverage percentage.

Fig. 3.11 and Fig. 3.12 show that both the nearest-neighbor-based scheme and the neighbor-number-based scheme are conservative. The actual coverage percentage they obtained is greater than the desired percentage, while the actual lifetime is shorter than the one obtained with SGP. The lifetime obtained with the probability-based algorithm is slightly longer than the one with SGP, but its coverage percentage fails to reach $P^*$.

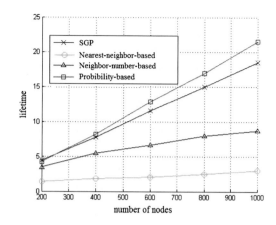

**FIGURE 3.12**

Network lifetime.

## 3.3 **Summary**

In this chapter, both PCCP and SGP can guarantee the desired coverage percentage. PCCP is a location-based scheme and SGP is location-free. Compared to complete coverage, percentage coverage requires fewer working nodes and hence prolongs the network lifetime.

For homogeneous WSNs, it is easier to generate Voronoi diagram to divide the area of interest. If sensor nodes have various sensing ranges, such a simple Voronoi diagram may not apply, instead, one kind of weighted Voronoi diagram may be used to assign the occupation areas for the sensor nodes (see [5,16]).

## **References**

[1] F. Aurenhammerv, Voronoi diagrams – a survey of a fundamental geometric data structure, ACM Computing Surveys 23 (3) (1991) 345–405.

[2] P. Bahl, V. Padmanabhan, RADAR: an in-building RF-based user location and tracking system, in: Proc. of Infocom'2000, 2, Tel Aviv, Israel, Mar. 2000, pp. 775–784.

[3] H. Bai, X. Chen, Y.C. Ho, X. Guan, Percentage coverage configuration in wireless sensor networks, in: ISPA 2005, in: LNCS, vol. 3758, 2005, pp. 780–791.

[4] H. Bai, X. Chen, B. Li, D. Han, A location-free algorithm of energy-efficient connected coverage for high density wireless sensor networks, Discrete Event Dynamic Systems 17 (1) (2007) 1–21.

[5] H. Bai, X. Chen, X. Guan, Preserving coverage for wireless sensor networks of nodes with various sensing ranges, in: Proceedings of 2006 IEEE International Conference on Networking, Sensing and Control, ICNSC'2006, Ft. Lauderdale, Florida, USA, April 23-25, 2006, pp. 54–59.

[6] N. Bulusu, J. Heidemann, D. Estrin, GPS-less low cost outdoor localization for very small devices, IEEE Personal Communications Magazine 7 (5) (2000) 28–34.

[7] Crossbow technology: wireless sensor networks, http://www.xbow.com/.

[8] S. Ganeriwal, R. Kumar, M.B. Srivastava, Timing-sync protocol for sensor networks, in: ACM SenSys'03, Los Angeles, CA, USA, November 2003, pp. 138–149.

[9] T. He, C. Huang, B.M. Blum, J.A. Stankovic, T. Abdelzaher, Range-free localization schemes for large scale sensor networks, in: MobiCom'03, San Diego, California, USA, September 2003.

[10] B. Hofmann-Wellenhof, H. Lichtenegger, J. Collins, Global Positioning System: Theory and Practice, 4th edition, Springer-Verlag, 1997.

[11] C. Hsin, M. Liu, Network coverage using low duty-cycled sensors: random & coordinated sleep algorithms, in: IPSN'04, Berkeley, California, USA, April 2004, pp. 433–442.

[12] J. Jiang, W.H. Dou, A coverage-preserving density control algorithm for wireless sensor networks, Lecture Notes in Computer Science 3158 (2004) 42–55.

[13] Q. Li, D. Rus, Global clock synchronization in sensor networks, in: Proceedings of the IEEE Infocom'04, Hong Kong, China, March 7–11, 2004, pp. 564–574.

[14] M. Maroti, P. Kusy, P. Simon, P. Ledeczi, The flooding time synchronization protocol, in: Proceedings of the 2nd International Conference on Embedded Networked Sensor Systems, 2004, pp. 39–49.

[15] D. Niculescu, B. Nath, DV based positioning in ad hoc networks, Telecommunication Systems 22 (1–4) (2003) 267–280.

[16] A. Okabe, B. Boots, K. Suguihara, S.N. Chiu, Spatial Tessellations: Concepts and Applications of Voronoi Diagrams, Wiley, 2000.

[17] V. Ramadurai, M.L. Sichitiu, Localization in wireless sensor networks: a probabilistic approach, in: Proceedings of the 2003 International Conference on Wireless Networks, ICWN'03, Las Vegas, NV, USA, June 23–26, 2003, pp. 275–281.

[18] A. Savvides, C.C. Han, M.B. Srivastava, Dynamic fine-grained localization in ad-hoc networks of sensors, in: MobiCom'01, Rome, Italy, July 2001, pp. 166–179.

[19] D. Tian, N.D. Georganas, A coverage-preserving node scheduling scheme for large wireless sensor network, in: WSNA'02, Atlanta, Georgia, USA, September 2002.

[20] D. Tian, N.D. Georganas, Location and calculation-free node-scheduling schemes in large wireless sensor networks, Ad Hoc Networks 2 (2004) 65–85.

[21] X. Wang, G. Xing, Y. Zhang, C. Lu, R. Pless, C. Gill, Integrated coverage and connectivity configuration in wireless sensor networks, in: ACM SenSys'03, Los Angeles, CA, USA, November 2003.

[22] F. Ye, G. Zhong, S. Lu, L. Zhang, PEAS: a robust energy conserving protocol for long-lived sensor networks, in: ICNP'02, Paris, France, November 2002, pp. 200–201.

[23] H. Zhang, J.C. Hou, On deriving the upper bound of alpha-lifetime for large sensor networks, in: MobiHoc'04, Roppongi, Japan, May 2004, pp. 121–132.

# Dynamic target detection

4

## CONTENTS

Wireless sensor networks which consist of a large number of densely deployed sensor nodes have a wide range of applications, such as military sensing, physical security, environment monitoring, traffic surveillance [1,7]. One example is the project ExScal, which is designed as a sensor network platform for reliable detection, classification, and quick reporting of rare, random, and ephemeral events. The project ExScal seeks to demonstrate that, being spread out over a 10 square kilometers area, a network composed of 10,000 sensors is capable of discriminating civilians, soldiers and vehicles [9].

When the environment of interest is inaccessible or located in a hostile area, sensor nodes may be air-dropped from an aircraft or in other ways, which results in a random placement [8]. For example, a number of wireless magnetic sensors are sprinkled along a road; these sensors automatically form a network once they hit the ground and then begin to scan the environment for magnetic signals. When a vehicle rolls by, from its magnetic signature, sensor nodes can recognize what kind of vehicle it is and estimate its speed and direction.

Wireless sensor nodes are usually equipped with a limited power source [1]. For example, as an off-the-shelf sensor model, MICA2 is able to work continuously for about one day with a battery capacity of 0.5Ah [13]. It may be impossible for sensor nodes to recharge their batteries. Hence power conservation and management are of great importance [1]. Power conservation protocol is usually applied to prolong the network lifetime. Although sensor nodes can negotiate to decide their working states by communication, randomly independent state switching schemes are also widely adopted (see, e.g., [10–12]) because such schemes are usually simple and free of communication.

Randomly Deployed Wireless Sensor Networks. https://doi.org/10.1016/B978-0-12-819624-3.00009-4

The task of this chapter is to design a WSN with given budget for sensor nodes, to monitor an area of interest for target detection with a predefined probability and keep the network working for a reasonable time. Both the detection probability and the network lifetime measure the quality of service of the network. We present the model of a randomly deployed WSN. Based on this model, by analysis, we get the detection probability as an explicit function of the network parameters and then optimize the quality of service of the network [6].

We consider a WSN in which all sensor nodes are identical in terms of energy, sensing radius, and any other function or capability. Sensor nodes are independently and identically distributed in the area of interest following a uniform distribution. Therefore the probability density function of each node located at point $a$ is

$$u(a) = \begin{cases} \dfrac{1}{S}, & a \text{ is in the area of interest;} \\ 0, & \text{otherwise,} \end{cases}$$

where $S$ is the total area of the area of interest. Here we assume that, compared to $S$, the sensor node is small enough to be viewed as a point. A target may enter the area of interest by crossing the border or it may drop in from the air. Hence any point in the area could be an entrance for targets.

Our basic idea is using the quantity (number of nodes) to trade for the quality (detection probability). We let each node be working/sleeping (or ON/OFF) randomly and independently. Sensor nodes can sense and communicate when working. But when sleeping, they neither sense nor communicate, hence consume no or little energy. Therefore, each node can live longer so that the network lifetime will be prolonged. Because of the random placement and the working/sleeping scheme for sensor nodes, the network is less likely to detect any target for sure. We define the detection probability as the probability that the network detects a target within its entire course of moving in the area of interest. Obviously, there is a conflict between power conservation and target detection. Sensor nodes not only need to schedule their working time so that they can live as long as the expected network lifetime, but also they should cooperate to guarantee a predefined detection probability.

Passive sensors such as magnetometer sensors and infrared sensors have sensing ranges. For the sake of convenience for analysis, we adopt a boolean sensing model, which means that a node's sensing range is a disc. In fact, the boolean sensing model is commonly applied in the study of sensing coverage (see e.g. [3,4,14]).

The assumptions imposed in this chapter are as follows.

**Assumption 4.1.** *The area of interest is a unit square.*

**Assumption 4.2.** *Sensor nodes are randomly and uniformly distributed in the area of interest.*

**Assumption 4.3.** *Sensor nodes are homogeneous.*

**Assumption 4.4.** *A boolean sensing model is adopted.*

**Assumption 4.5.** *The total energy of a sensor node cannot afford it to be working for the expected network lifetime.*

Generally, the area of interest may be of various shapes. For the sake of convenience for the analysis, it is assumed as a unit square. The results based on this assumption can be easily applied to different shapes.

## 4.1 State switching scheme

In the network, time synchronization is unnecessary for sensor nodes. Each of them takes a working or a sleeping state with probability $\rho$ or $1 - \rho$. In addition, each node must keep the state for a time period $T$ before the next period starts. We call $\rho$ the working probability. The state switching scheme combines randomized activation and duty-cycle, the pair $(\rho, T)$ essentially determines the behavior of each node or even the network at large. After setup, each node will follow this scheme independently with the same value of $(\rho, T)$.

Denote by $\{\Omega_m, m \geqslant 0\}$ the sequence of one node's states, where $m$ represents its $m$th $T$. Hence we have

$$\text{Prob}(\Omega_m = \text{working}) = \rho, \quad \text{Prob}(\Omega_m = \text{sleeping}) = 1 - \rho, \ m \geqslant 0.$$

Moreover, the probability for the node to keep working (or sleeping) for $kT$ follows a geometric distribution. Therefore, instead of determining its state at the beginning of each period, the node only needs to take the following steps for scheduling.

Step 1: When working, the node generates a sequence of random numbers, $w_1$, $w_2, \cdots, w_k$, where $w_i \sim U(0, 1)$ for any $1 \leq i \leq k$; $w_k$ is the first number satisfying $w_k < \rho$, namely $w_i \geq \rho$ for any $1 \leq i < k$.

Step 2: If $k = 1$, the node will stay working in the following $T$ and repeat Step 1.

Step 3: If $k > 1$, the node will keep sleeping in the following $(k-1)T$, then switch to working for the $k$th $T$ and repeat Step 1.

## 4.2 Analysis of detection probability

Compared to the area of interest, the target's size is assumed to be small enough to be ignored, hence we view it as a point. The sensing radius of every sensor node is $r$. When a target is moving in the area of interest, only the nodes which are close to its path may possibly sense it. For simplicity, we assume that once the node is working and a target appears in its sensing range, then the node can detect the target immediately. Now we analyze the detection probability of a node, which implements the state switching scheme described in the above section.

## 4.2.1 Detection probability of an individual sensor

Suppose both working probability $\rho$ and period $T$ are given. Denote the time for a target getting through the sensing range of a node as $\Delta t$. Moreover, suppose $k$ is a nonnegative integer satisfying $k = \lfloor \Delta t / T \rfloor$, where $\lfloor \cdot \rfloor$ is the floor function; then $\Delta t / T - 1 < k \leqslant \Delta t / T$, namely $kT \leqslant \Delta t < (k+1)T$.

Two time axes shown in Fig. 4.1 illustrate the possible cases which may occur between $\Delta t$ and a node's state switching. The time length between two vertical lines equals $T$. $\Delta t$ could start from any point in $[0, T)$. If the starting point falls in $[0, (k+1)T - \Delta t]$, $\Delta t$ will end in the $(k+1)$th $T$ as shown in Fig. 4.1B; if the starting point falls in $((k+1)T - \Delta t, T)$ as shown in Fig. 4.1A, then $\Delta t$ will end in the $(k+2)$th $T$.

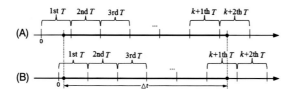

**FIGURE 4.1**

Illustration of the two possible cases of $\Delta t$ and $T$.

Each sensor node independently switches its state without time synchronization. Hence, as a target getting through a node's sensing range, the probabilities for the two cases shown in Fig. 4.1A and Fig. 4.1B are $(\Delta t - kT)/T$ and $((k+1)T - \Delta t)/T$, respectively. Within every $T$ a sensor node has probability $1 - \rho$ to be sleeping, so the probability that the node cannot detect the target is

$$
\begin{aligned}
q(\Delta t) &= (1-\rho)^{k+2}(\Delta t - kT)/T + (1-\rho)^{k+1}((k+1)T - \Delta t)/T \\
&= (1-\rho)^{k+1}(1 - \rho(\Delta t / T - k)).
\end{aligned}
\tag{4.1}
$$

Suppose a target's velocity is $v$ and it is moving along a line. Sensor nodes which are far from the target's path cannot detect it. There is a detection zone along its path as shown in Fig. 4.2. Since the perpendicular distance between the target's path and the sensor is $x$, the length of the target's path within the sensing range of the node is $2\sqrt{r^2 - x^2}$. So we have $\Delta t = 2\sqrt{r^2 - x^2}/v$.

Since sensor nodes are randomly and uniformly distributed; $x$ follows a uniform distribution, namely $x \sim U(0, r)$. With $\Delta t = 2\sqrt{r^2 - x^2}/v$, the mean of $q(\Delta t)$ is

$$
\begin{aligned}
\bar{q} &= E_{\Delta t}\{q(\Delta t)\} \\
&= E_x \left\{ (1-\rho)^{\lfloor \frac{\Delta t}{T} \rfloor + 1} \left[ 1 - \rho \left( \frac{\Delta t}{T} - \left\lfloor \frac{\Delta t}{T} \right\rfloor \right) \right] \right\}
\end{aligned}
$$

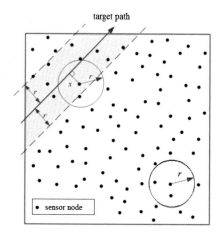

target path

• sensor node

**FIGURE 4.2**

Illustration of detection zone and sensing range of a node.

$$= (1-\rho) \int_0^r (1-\rho)^{\left\lfloor \frac{2\sqrt{r^2-x^2}}{vT} \right\rfloor} \left[ 1 - \rho \left( \frac{2\sqrt{r^2-x^2}}{vT} - \left\lfloor \frac{2\sqrt{r^2-x^2}}{vT} \right\rfloor \right) \right] \frac{dx}{r}$$

$$= (1-\rho) \int_0^1 (1-\rho)^{\left\lfloor \frac{2r\sqrt{1-z^2}}{vT} \right\rfloor} \left[ 1 - \rho \left( \frac{2r\sqrt{1-z^2}}{vT} - \left\lfloor \frac{2r\sqrt{1-z^2}}{vT} \right\rfloor \right) \right] dz.$$

$$(4.2)$$

On average, any node in the detection zone can detect the target with probability

$$\bar{p} = 1 - \bar{q}. \tag{4.3}$$

Both $\bar{p}$ and $\bar{q}$ are functions of $\rho$, $T$, $r$ and $v$. So we rewrite $\bar{p}$ and $\bar{q}$ as

$$\bar{p}(\rho, T, r, v) = 1 - \bar{q}(\rho, T, r, v),$$

$$\bar{q}(\rho, T, r, v) = (1-\rho) \int_0^1 (1-\rho)^{\left\lfloor \frac{2r\sqrt{1-z^2}}{vT} \right\rfloor}$$

$$\times \left[ 1 - \rho \left( \frac{2r\sqrt{1-z^2}}{vT} - \left\lfloor \frac{2r\sqrt{1-z^2}}{vT} \right\rfloor \right) \right] dz. \tag{4.4}$$

With Eq. (4.4), we know that a bigger $\rho$ leads to a bigger detection probability. When $\rho = 1$, which means that the sensor nodes stay working without sleeping, then $\bar{p} = 1$, which means that the nodes certainly will detect the target. On the contrary, $\rho = 0$ means that the nodes are always in the sleeping state; therefore $\bar{p} = 0$. A network needs a smaller $\rho$ to have a longer lifetime. If $\rho$ is close to 0, $\bar{p}$ will go to 0 and the sensor nodes hardly detect targets. If a network is expected to have a bigger detection probability, it should have more nodes, that is, one trades quantity for quality.

In addition, smaller $v$ and/or shorter $T$ can also make a bigger detection probability.

**Proposition 4.1.** *The mean of detection probability by an individual sensor node, $\bar{p}$, defined in Eq. (4.4), is monotonously increasing as $T$ is decreasing or $v$ is decreasing.*

The proof of Proposition 4.1 can be found in [6]. Here we provide an intuitive explanation of this proposition. A slower target needs more time to get through a node's sensing range, hence it is more likely to be detected. On the other hand, even the target's speed is fixed, if the period $T$ is shorter, the sensor nodes will switch state more frequently and then they have more chance to be in the working state to detect the target.

When $v \to 0$, the target is essentially stationary. Any node in the detection zone can detect the target and therefore $\bar{p} \to 1$. When $v \to \infty$, the target rapidly moves through the area of interest, and any node in the detection zone has no chance to switch state. Hence the detection probability equals the probability of being working, that is, $\bar{p} \to \rho$.

Suppose $0 < v \leqslant \hat{v}$, i.e. $\hat{v}$ is an upper bound of $v$. With Proposition 4.1, $\bar{p}(\rho, T, r, \hat{v}) \leq \bar{p}(\rho, T, r, v)$ and the former one is a conservative estimate of the latter one. If a target's speed is not a constant, on average, $\bar{p}(\rho, T, r, \hat{v})$ is the lower bound of the detection probability for any node in the detection zone.

A smaller $T$ results in a bigger $\bar{p}$. Hence $T$ should be as small as possible to maximize $\bar{p}$. However, due to the physical property of sensor nodes, $T$ cannot be infinitely small. Sensor nodes need some time to warm up (set up) when switching from sleeping to working.

Similarly, a bigger $r$ results in a bigger $\bar{p}$. But the sensing radius $r$ usually is determined by the physical properties. Hence we assume $r$ is fixed for a real sensor model.

### 4.2.2 Detection probability with given target's path

When a target traverses the area of interest along one line, a certain number of nodes may detect it. As shown in Fig. 4.2, all nodes within the detection zone have a chance to detect the target. In the following, we analyze the detection probability when a target is getting through along a given path with a constant speed.

**Proposition 4.2.** *Suppose one target moves in the area of interest along a line with speed $v$ and its path's length is $L$. On average, the probability that the network detects the target is*

$$\bar{p}_{\mathrm{L}}(L, \rho, T, r, v, N) = 1 - (1 - 2rL\bar{p})^N, \tag{4.5}$$

*where $N$ is total number of nodes in the network, $r$ is the sensing radius, $\rho$ and $T$ are the parameters for the state switching scheme; $\bar{p}$ is defined by Eq. (4.4).*

*Proof.* Since the width of detection zone is $2r$, the area of the detection zone for the target is $2rL$.

If there are $i$ nodes, say $O_1, O_2, \cdots, O_i$, located in the detection zone, then the network will fail to detect the target only when each of these $i$ nodes fails. Since each node runs independently, we get

$$\text{Prob(All of these } i \text{ nodes fail in detection)} = \prod_{j=1}^{i} \text{Prob}( O_j \text{ fails in detection)}.$$

Assumption 4.1 assumes that the area of interest is a unit square. When $N$ nodes are uniformly distributed in the area of interest, the integer $i$ obeys the binomial distribution $B(N, 2rL)$, that is,

$$
\begin{aligned}
p_{\mathrm{L}}(L, \rho, T, r, v, N) &= 1 - \sum_{i=0}^{N} \left[ \binom{N}{i} (1 - 2rL)^{N-i} (2rL)^i \right. \\
&\quad \left. \times \text{Prob(All of these } i \text{ nodes fail in detection)} \right] \\
&= 1 - \sum_{i=0}^{N} \left[ \binom{N}{i} (1 - 2rL)^{N-i} (2rL)^i \right. \\
&\quad \left. \times \prod_{j=1}^{i} \text{Prob}(O_j \text{ fails in detection)} \right].
\end{aligned}
$$

Due to the independence of the node's location, on average, we have

$$
\begin{aligned}
\bar{p}_{\mathrm{L}}(L, \rho, T, r, v, N) &= \mathrm{E}[p_{\mathrm{L}}(L, \rho, U, r, v, N)] \\
&= 1 - \sum_{i=0}^{N} \left[ \binom{N}{i} (2rL)^i (1 - 2rL)^{N-i} \right. \\
&\quad \left. \times \mathrm{E}\left( \prod_{j=1}^{i} \text{Prob}(O_j \text{ fails in detection)} \right) \right] \\
&= 1 - \sum_{i=0}^{N} \left[ \binom{N}{i} (2rL)^i (1 - 2rL)^{N-i} (\bar{q})^i \right] \\
&= 1 - (1 - 2rL + 2rL\bar{q})^N \\
&= 1 - (1 - 2rL\bar{p})^N. \quad \square
\end{aligned}
$$

Proposition 4.2 shows that a longer $L$ leads to a bigger detection probability as there are more nodes located in the detection zone.

In Eq. (4.5), $2rL$ is the area of detection zone. Suppose the area of the detection zone is $s$. By replacing $2rL$ with $s$, we have a more general form of Eq. (4.5), namely

$$\bar{p}_{\mathrm{s}}(s, \rho, T, r, v, N) = 1 - (1 - s\bar{p})^N. \tag{4.6}$$

Eq. (4.6) has wide applicability. For example, a target may take a curve path in the area of interest. In this case, the target's detection zone may be calculated by using several line segments to approximate the curve. Another scenario is that a target lands on the area of interest and moves around inside, where Eq. (4.6) works as well.

In addition, if one more sensor is added in the network, its contribution to the detection probability is

$$
\begin{aligned}
\Delta p_s &= \bar{p}_s(s, \rho, T, r, v, N+1) - \bar{p}_s(s, \rho, T, r, v, N) \\
&= (1 - s\bar{p})^N - (1 - s\bar{p})^{N+1} \\
&= s\bar{p}(1 - s\bar{p})^N.
\end{aligned}
$$

$\Delta p_s$ is getting smaller as $N$ is getting bigger. That is, when the network has already a large number of nodes, one more or one less node can hardly affect detection probability.

When the desired detection probability is $P^*$, with Eq. (4.6), the number of nodes required by the network is

$$
N = \frac{\ln(1 - P^*)}{\ln(1 - s\bar{p})}.
$$

Denote the total time for a target moving within the area of interest by $\bar{t}$. If its path length is $L$ and speed is $v$, then $\bar{t} = L/v$. Denote $\bar{p}_L(L, \rho, T, r, v, N, t)$ as the average probability for the target to be detected before $t$, where $0 < t < \bar{t}$, then we have

$$
\bar{p}_L(L, \rho, T, r, v, N, t) \approx 1 - (1 - 2rvt\bar{p})^N. \tag{4.7}
$$

Eq. (4.7) reveals the relationship between detection probability and the time by the target spends in the area of interest. It helps us to understand the behavior of the network in target detection.

Conversely, if the network is required to detect the target with probability $P^*$ before $t$ after it enters, Eq. (4.7) also indicates the condition of the corresponding $N$ for the network.

If a target's speed changes with time, by using an upper bound of its speed, we get a conservative estimate of the detection probability.

## 4.2.3 Number of nodes which success in detecting target

We usually wonder how many sensor nodes detect a target when it is getting through the area of interest. When more nodes are detecting the target, locating and tracking become more easier with more reliable information about the target.

**Proposition 4.3.** *Suppose the area of a target's detection zone is s, on average, the probability that there are exactly n sensors detecting the target is*

$$
\bar{p}_D(n, \rho, T, r, v, s, N) = \binom{N}{n} (s\bar{p})^n (1 - s\bar{p})^{N-n}.
$$

*Proof.* Suppose there are $i$ nodes, i.e. $O_1, O_2, \cdots, O_i$, located in the detection zone, where $i \geq n$. The probability that $n$ out of $i$ nodes detect the target and the rest $i - n$ nodes fail in detection is

$$b(n, i) = \binom{i}{n} \prod_{j=1}^{n} \text{Prob}(O_j \text{ detects the target}) \prod_{j=n+1}^{i} \text{Prob}(O_j \text{ fails in detection}).$$

Taking the expectation with respect to each node's location, we have

$$E[b(n, i)] = \binom{i}{n} (\bar{p})^n (\bar{q})^{i-n}.$$

Since $N$ nodes are distributed in the area of interest following a uniform distribution, $i$ obeys the binomial distribution $B(N, s)$. The probability that there are exactly $n$ nodes detecting the target is

$$p_D(n, \rho, T, r, v, s, N) = \sum_{i=n}^{N} \binom{N}{i} s^i (1 - s)^{N-i} b(n, i).$$

Taking the average on both sides and using the formula $\binom{N}{i}\binom{i}{n} = \binom{N}{n}\binom{N-n}{i-n}$, we have

$$
\begin{aligned}
\bar{p}_D(n, \rho, T, r, v, s, N) &= \sum_{i=n}^{N} \left[ \binom{N}{i} s^i (1-s)^{N-i} \binom{i}{n} (\bar{p})^n (\bar{q})^{i-n} \right] \\
&= \sum_{i=n}^{N} \left[ \binom{N}{n} \binom{N-n}{i-n} s^i (1-s)^{N-i} (\bar{p})^n (\bar{q})^{i-n} \right] \\
&= \binom{N}{n} (s\bar{p})^n \sum_{i=n}^{N} \left[ \binom{N-n}{i-n} (1-s)^{N-i} (s\bar{q})^{i-n} \right] \\
&= \binom{N}{n} (s\bar{p})^n (1-s+s\bar{q})^{N-n} \\
&= \binom{N}{n} (s\bar{p})^n (1-s\bar{p})^{N-n}. \quad \square
\end{aligned}
$$

## 4.3 **Performance optimization**

Denote by $N_L(\rho, T, r, v, L^*)$ the number of nodes to guarantee the detection probability, $P^*$, when a target's speed is $v$ and path length is $L^*$. With Eq. (4.5), we get

$$N_L(\rho, T, r, v, L^*) = \frac{\ln(1 - P^*)}{\ln(1 - 2L^* r \bar{p})}, \tag{4.8}$$

where $\bar{p}$ is defined by Eq. (4.4).

From Eq. (4.4) and Eq. (4.8), we see that increasing $r$ not only leads to $\bar{p}$ increasing but also enlarges the area of a node's detection area. Compared to increasing $\rho$ or decreasing $T$, increasing $r$ reduces the number of nodes more effectively. But for a passive sensor node, $r$ is mainly determined by its physical properties instead of being adjustable.

Different from the sensing radius, $(\rho, T)$ is the parameter pair in the state switching scheme. Choosing a proper value for $(\rho, T)$ to maximize $\bar{p}$ may reduce the number of nodes and pave the way for network performance optimization.

With Proposition 4.1, we know that $\bar{p}$ will increase as $T$ decreases. However, $T$ cannot be too small since a real sensor needs some time to warm up when switching from sleeping to working. Therefore we have $T \geq \underline{T} > 0$, where $\underline{T}$ is a lower bound of $T$.

On the other hand, $T$ is limited by the total energy of a sensor node. Denote by $L_{sn}$ the network lifetime. Suppose each sensor node has total energy $E_0$, consumes $Q_{on}$ per time unit when working and spends extra energy $e_s$ to switch state from sleeping to working. With Assumption 4.5, we have $L_{sn} > E_0/Q_{on}$. The working probability $\rho$ should satisfy $\rho Q_{on} L_{sn} < E_0$. To make the network live for $L_{sn}$, we should have

$$0 < \rho < \frac{E_0}{Q_{on} L_{sn}}.$$

The average energy consumption per node within every period is

$$E(\rho, T) = \rho T Q_{on} + \rho(1 - \rho)e_s.$$

To let a node live for $L_{sn}$, we also should have

$$\frac{E(\rho, T)}{T} L_{sn} \leqslant E_0, \tag{4.9}$$

that is,

$$\frac{\rho T Q_{on} + \rho(1 - \rho)e_s}{T} L_{sn} \leqslant E_0, \tag{4.10}$$

or

$$T \geq \frac{\rho(1 - \rho)e_s L_{sn}}{E_0 - \rho Q_{on} L_{sn}}. \tag{4.11}$$

### 4.3.1 Maximizing detection probability with given network lifetime

Since $\bar{p}$ will become bigger when $T$ is smaller, with Eq. (4.11) we have

$$T = \max\left\{\underline{T}, \frac{\rho(1 - \rho)e_s L_{sn}}{E_0 - \rho Q_{on} L_{sn}}\right\}, \tag{4.12}$$

where $e_s$, $E_o$ and $\underline{T}$ are assumed to be known. When $L_{sn}$ is given, $\bar{p}$ merely depends on the value of $\rho$.

The optimization problem is formulated as follows:

$$\max_{\rho} \bar{p}(\rho, T, r, v)$$

$$= 1 - (1 - \rho) \int_0^1 (1 - \rho)^{\left\lfloor \frac{2r\sqrt{1-z^2}}{v \cdot T} \right\rfloor} \left[ 1 - \rho \left( \frac{2r\sqrt{1-z^2}}{v \cdot T} - \left\lfloor \frac{2r\sqrt{1-z^2}}{v \cdot T} \right\rfloor \right) \right] dz$$

$$\begin{cases} T = \max \left\{ \underline{T}, \dfrac{\rho(1-\rho)e_s L_{sn}}{E_o - \rho Q_{on} L_{sn}} \right\}, \\ 0 < \rho < \dfrac{E_o}{Q_{on} L_{sn}}. \end{cases}$$

(4.13)

The second term in max for $T$ hinges on $\rho$. By setting

$$V(\rho) = \frac{\rho(1-\rho)e_s L_{sn}}{E_o - \rho Q_{on} L_{sn}},$$

(4.14)

and taking derivative on both sides, we have

$$\frac{dV(\rho)}{d\rho} = \frac{\left( E_o(1-\rho)^2 + \rho^2(Q_{on}L_{sn} - E_o) \right) e_s L_{sn}}{(E_o - \rho Q_{on}L_{sn})^2}.$$

(4.15)

Since $L_{sn} > E_o/Q_{on}$, we have $\frac{dV(\rho)}{d\rho} > 0$, that is, as $\rho$ increases $V(\rho)$ monotonously increases. With Eq. (4.12), when $V(\rho) > \underline{T}$, then $T = V(\rho)$; when $\rho$ is small enough such that $V(\rho) \leqslant \underline{T}$, then $T = \underline{T}$.

The following algorithm can solve Problem (4.13) numerically.

**Algorithm 4.1.** Optimizing $(\rho, T)$ with given $L_{sn}$
   Initialization;
   $\rho_0$ satisfies $V(\rho_0) = \underline{T}$ ($\rho_0$ is the minimum of $\rho$)
   $\bar{\rho} = E_o/(Q_{on}L_{sn})$ ($\bar{\rho}$ is an upper bound of $\rho$)
   $\Delta\rho = (\bar{\rho} - \rho_0)/J$, where $J$ is a large integer;
   $\rho^* = \rho_0$, $T^* = \underline{T}$, $\bar{p} = \bar{p}(\rho_0, \underline{T}, r, v)$;
   $j = 0$;
   Repeat (*searching for $\rho^*$ which maximize $\bar{p}$*)
           $j = j + 1$, $\rho_j = \rho_0 + j\Delta\rho$, $V_j = V(\rho_j)$ with Eq. (4.14),
           Calculate $pp = \bar{p}(\rho_j, V_j, r, v)$, with Eq. (4.4)
           if $pp > \bar{p}$, then
                   $\rho^* = \rho_j$, $T^* = V_j$, $\bar{p} = pp$,
           end if
   until $j = J$.

The solution of Problem (4.13), $(\rho^*, T^*)$, is a function of $L_{sn}$. Hence we denote it by $(\rho^*(L_{sn}), T^*(L_{sn}))$. With $(\rho^*(L_{sn}), T^*(L_{sn}))$, the total number of nodes can be obtained to maximize the network lifetime, that is,

$$M(L_{sn}) = \frac{\ln(1 - P^*)}{\ln(1 - 2L^* r \bar{p}(\rho^*(L_{sn}), T^*(L_{sn}), r, v))}, \tag{4.16}$$

where $L^*$ is the shortest path length for a target to traverse the monitoring area.

To secure the detection probability $P^*$, a larger $L_{sn}$ results in a smaller $\bar{p}$ and in return the network needs more sensor nodes.

### 4.3.2 Maximizing network lifetime with budget limit

Now we consider how to maximize the network lifetime, $L_{sn}$, with given budget on the sensor nodes. At the stage of system design, there is always a budget limit. Suppose $\bar{B}$ is the budget and $c$ is the cost per node. The following is the problem formulation of maximizing $L_{sn}$ subject to the budget limit:

$$\begin{aligned} &\max L_{sn} \\ &c M(L_{sn}) \leqslant \bar{B}. \end{aligned} \tag{4.17}$$

Problem (4.17) and Problem (4.13) together form a non-cooperative game, which is known as a Stackelberg problem with $L_{sn}$ as leader and $(\rho, T)$ as follower. The solution for Problem (4.13), $(\rho^*(L_{sn}), T^*(L_{sn}))$, is the response of $(\rho, T)$ to $L_{sn}$ [2].

The solution to Problem (4.13) and Problem (4.17) is the equilibrium $(\rho^*, T^*, L_{sn}^*)$ of the game. The following algorithm can find the equilibrium.

**Algorithm 4.2.** Finding the equilibrium $(\rho^*, T^*, L_{sn}^*)$
  Initialization;
  $L_u = E_0/Q_{on}$ ($L_u$ is the minimum of $L_{sn}$);
  $\Delta L_u =$ a unit of time (*e.g., one hour or one day*);
  $L_{sn} = L_u$;
  Repeat (*searching for the maximum of $L_{sn}$*)
            With Algorithm 4.1, find solution $(\rho^*(L_{sn}), T^*(L_{sn}))$
            Calculate $M(L_{sn})$ with Eq. (4.16)
            $L_{sn} = L_{sn} + \Delta L_u$
  until $\bar{B} < c M(L_{sn})$;
  $L_{sn}^* = L_{sn} - \Delta L_u$.

In this section, we formulate network performance optimization with a two-level model. At the lower level, parameter $\rho$ and $T$ are optimized subject to the energy constraint. At the upper level, the network lifetime $L_{sn}$ is maximized subject to the budget. In this way, we obtain a numerical equilibrium solution of Problem (4.17).

**Example 4.1** (Border surveillance). WSNs have been playing a major role in monitoring the area along the border between two countries.

Fig. 4.3 depicts a simple border surveillance system, where wireless sensor nodes are deployed on one side of the border to detect illegal crossing. Nodes are deployed within one band. Without loss of generality, we suppose a small segment of border is straight and the width of the surveillance band is fixed. Suppose each node adopts the state switching scheme presented in Section 4.1.

Denote the maximum velocity of illegal crossing as $\hat{v}$, the shortest track length as $L^*$.

As shown in Fig. 4.3, suppose the width of the border between two countries is 1000 meters. We consider one segment of the border. Its length is also 1000 meters. Hence the area of monitoring is $1000 \times 1000$ square meters.

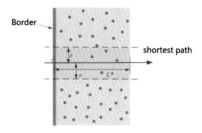

**FIGURE 4.3**

Border surveillance.

Suppose the budget for sensor nodes is \$5000, namely $\bar{B} = 5000$. There are two types of sensors, which share common $T$ and $e_s$. $T = 1$ s, $e_s = 0.005$ J. The other parameters are different, as shown in Table 4.1. Suppose a vehicle's maximal speed

**Table 4.1** Parameters of sensor nodes.

| Type | $E_o$ | $Q_{on}$ | $r$ | $c$ |
|---|---|---|---|---|
| A | 15000 J | 0.05 W | 20 m | \$ 10 |
| B | 10000 J | 0.025 W | 15 m | \$ 6 |

is $\hat{v} = 5$ m/s. The network is required to detect such a vehicle with probability 0.99, namely $P^* = 0.99$.

For type A and type B sensor nodes, we solve Problem (4.17) with Algorithm 4.2, respectively. For type A, the network lifetime is 88 days with $(\rho^*, T^*) = (0.0358, 1s)$. However, for type B, the network lifetime is 80 days. Therefore we should choose type A for the surveillance system.

## 4.4 **Summary**

In this chapter, a model of randomly deployed WSN for dynamic-target detection is presented. Since the area of interest is assumed to be a unit square, the number of sensor nodes required by the network may serve as the node density, which may be applicable to an area of interest of any shape. The state switching scheme lets the sensor nodes work independently without coordination so that they save much energy on communication. The state switching scheme provides the network with a good scalability as the number of nodes increases. Analysis and results deliver deep insights and useful information for the design of randomly deployed WSNs.

As sensor nodes switch their states between working and sleeping, the topology of the network varies frequently. This feature leads to difficulty to routing for data transmission. Even though there exist a few data transmission protocols for large scale WSNs (see, e.g., [5,15]), a randomly deployed WSN demands data transmission protocols to correctly be incorporated within the state switching scheme. A robust and flexible protocol for data transmission will be presented in Chapter 5.

# References

[1] I.F. Akyildiz, W. Su, Y. Sankarasubramaniam, E. Cayirci, A survey on sensor networks, IEEE Communications Magazine (August 2002) 102–114.
[2] J.P. Aubin, Optima and Equilibria: An Introduction to Nonlinear Analysis, Springer, Berlin, 1993.
[3] M. Cardei, J. Wu, Handbook of Sensor Networks, CRC Press, Boca Raton, 2002.
[4] K. Chakrabary, S.S. Iyengar, H. Qi, E. Cho, Grid coverage for surveillance and target location in distributed sensor networks, IEEE Transactions on Computer 51 (2002) 1448–1453.
[5] I. Chatzigiannakis, S. Nikoletseas, P. Spirakis, Efficient and robust protocols for local detection and propagation in smart dust networks, Mobile Networks and Applications 10 (2005) 133–149.
[6] X. Chen, H. Bai, Y.C. Ho, Design of a randomly distributed sensor network for target detection, Automatica 43 (10) (2007) 1713–1722.
[7] C. Chong, S.P. Kumar, Sensor networks: evolution, opportunities and challenges, Proceeding of the IEEE 91 (8) (Aug. 2003).
[8] T. Clouqueur, V. Phipatanasuphorn, P. Ramanathan, K.K. Saluja, Sensor deployment strategy for detection of targets traversing a region, ACM Mobile Networks and Applications 8 (2003) 453–461.
[9] P. Dutta, M. Grimmer, A. Arora, S. Bibyk, D. Culler, Design of a wireless sensor network platform for detecting rare, random, and ephemeral events, in: Proceedings of the 4th International Symposium on Information Processing in Sensor Networks, Los Angeles, CA, 2005.
[10] C. Gui, P. Mohapatra, Power conservation and quality of surveillance in target tracking sensor networks, in: MobiCom 2004, Philadelphia, PA, 2004.
[11] S. Pattem, S. Poduri, B. Krishnamachari, Energy-quality tradeoffs for target tracking in wireless sensor networks, in: Second International Workshop on Information Processing in Sensor Networks, 2003.
[12] S. Ren, Q. Li, H. Wang, X. Chen, X. Zhang, Analyzing object detection quality under probabilistic coverage in sensor networks, in: Proceedings of the 13th International Workshop on Quality of Service, IWQoS'05, in: Lecture Notes in Computer Science, vol. 3552, Springer, Berlin, 2005, pp. 107–122.
[13] A. Rev, MTS/MDA Sensor and data acquisition boards user's manual, Document 7430-0020-02, Crossbow Technology, 2003.
[14] S. Slijepcevic, M. Potkonjak, Power efficient organization of wireless sensor networks, in: IEEE International Conference on Communication, Helsinki, Finland, 2001, pp. 472–476.
[15] M. Zorzi, R.R. Rao, Geographic random forwarding (GeRaF) for ad hoc and sensor networks: energy and latency performance, IEEE Transactions on Mobile Computing 2 (2003) 337–348.

# Probabilistic forwarding protocols

5

## CONTENTS

Data transmission has been an important research topic in WSNs. By radio frequency communication, the energy consumption is proportional to the $n$th ($2 \leq n \leq 4$) power of the transmission distance. To save energy, short distance multiple hop communication becomes preferable to long distance direct communication. Data is transmitted from its source node to the base station or the sink node via relaying by intermediate sensor nodes. Usually there exist multiple routes from the source to the base station. Routing is to find the proper one for data transmission.

Data transmission protocols are based on the topological structure of networks. In general, topological structures of WSNs are classified as cluster topology, tree topology and gradient topology, as shown in Fig. 5.1.

In a cluster topology (see Fig. 5.1A), nodes are divided into different clusters (dotted ovals). In each cluster, all data first converge to the cluster head (gray circle). Cluster heads usually directly communicate with the base station. If one cluster head shows failure, then data from the rest of the nodes in the cluster will be unable to reach the base station. Cluster head switching can efficiently balance energy consumption in the network. However, head switching may need extra communication for negotiation among nodes.

In a tree topology (see Fig. 5.1B), each node in a hierarchy level has point-to-point links with each adjacent node on its below level. The base station is at the top level (or

**Randomly Deployed Wireless Sensor Networks. https://doi.org/10.1016/B978-0-12-819624-3.00010-0**
Copyright © 2020 Tsinghua University Press. Published by Elsevier Inc. All rights reserved.

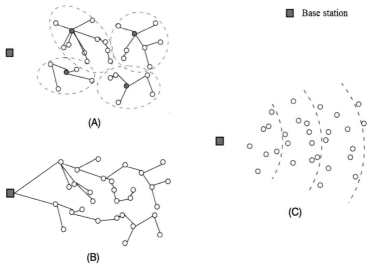

Base station

(A)

(B)

(C)

**FIGURE 5.1**

Three kinds of topological structure. (A) Cluster topology. (B) Tree topology. (C) Gradient topology.

root). Data from any node in the tree needs to be relayed by the intermediate nodes on the route from the node to the base station. If one intermediate node shows failure, all data from the nodes which are located at its lower levels cannot reach the base station. Tree structure also results in uneven energy consumption in the network.

In a gradient topology (see Fig. 5.1C), all nodes create their gradients based on the minimal number of hops to the base station. Data travels along the gradient descent direction and finally converges to the base station. The gradient topology can support more flexible and multi-path routing protocols, which may spend more energy on transmission but enhance the reliability of data reaching the base station.

Data transmission protocols in WSNs may be roughly classified into two categories according to the way of routing. In one category, the transmission route is predetermined for packets and intermediate nodes have to relay packets from the source to the base station. Route tables are determined by the topology of the network and need to be updated when the topology changes. For example, LEACH [8], as a clustering-based protocol, lets each node maintain its route table by communicating with neighboring nodes. In the other category, packets are transmitted without any specific route. Flooding is such a kind of protocol, and so is Gossiping. They do not require a packet to follow any specific route to the base station; hence no route table is necessary any more.

Haas et al. [7] develop a gossiping-based approach, where each node forwards a packet with some probability. Combined with variations of flooding, the gossiping-based approach can reduce the overhead of routing protocols. This work first introduces forward probability into routing protocols.

Wu et al. [14] present a selective forwarding probability. Neighbors as the next hop are selected with some probability. The selective forwarding probability takes into account of the node degree and link loss in order to enhance the reliability of selective forwarding.

Barrett et al. [3] propose a family of light-weight and robust multi-path routing protocols in which an intermediate sensor node forwards a message with a probability which depends on various parameters, such as the distance of the source node to the destination, the distance of the node to the destination, or the number of hops a packet has already traveled.

In this chapter, a probabilistic forwarding approach for data transmission is presented. With this approach, each message may not reach the base station with certainty but with a predefined success probability. There are many reasons for such a relaxation. First of all, in a large scale WSN, data from different sensor nodes may be similar or duplicated, the absence of any small part of data can hardly affect the network performance. Secondly, in some real applications, a high success probability is acceptable and good enough. Thirdly, the cost of demanding absolute certainty is usually unaffordable.

Notice that in [3,7,14], all probabilistic approaches are used in a heuristic manner. In this chapter, we analyze and obtain the condition for the relay probability to meet the requirement on data transmission.

## 5.1 Probabilistic forwarding (ProFor)

We study the way of data transmission and develop a probabilistic forwarding protocol. This protocol lets relay nodes transmit packets with a certain probability to let messages reach the base station with a predefined probability [5].

### 5.1.1 Model description

The following assumptions are imposed in this chapter.

**Assumption 5.1.** *Sensor nodes are randomly and densely deployed in the area of interest.*

**Assumption 5.2.** *Each sensor node knows its gradient.*

**Assumption 5.3.** *Each sensor node has an unique ID.*

**Assumption 5.4.** *Each sensor node adopts the communication radius $R_0$.*

In addition, some definitions on nodes and messages are as follows.

**Definition 5.1.** *A sensor node is called an h-hop node if its gradient is h.*

**Definition 5.2.** *A message is called an h-hop message if its source is an h-hop node.*

**Definition 5.3.** *If an $(h-1)$-hop node and an $h$-hop node can communicate with communication radius $R_0$, then the former one is called a 1-hop downstream neighbor of the latter one; and the latter one is called a 1-hop upstream neighbor of the former one.*

**Definition 5.4.** *If different $(h-1)$-hop nodes have a common $h$-hop node as their 1-hop upstream node, then they are called sibling nodes with respect to the $h$-hop node.*

### Generating the gradient for each node

Suppose the network has only one stationary base station and it is located at the border or inside the area of interest and all nodes would not move after the deployment.

The network performs an initialization to generate the gradient for each node by following a similar way to [11]: the base station initiates a gradient by sending its neighboring sensor nodes a message with communication radius $R_0$. The message includes a count set to 1. Each node which receives the message remembers the value of the count and forwards the message to its neighbors with communication radius $R_0$ but with the count incremented by 1. Therefore, a wave of messages propagates outwards from the base station. Each node keeps the minimum count value it has received and ignores messages which contain larger count values. A node, say $O_j$, gets the minimum hop count, say $h_j$. $h_j$ represents the length of shortest path from $O_j$ to the base station in terms of communication hops. $h_j$ is also called the gradient of $O_j$ with respect to the base station.

## 5.1.2 Analysis of relay probability

In this chapter, messages and packets are two related but different concepts. A message refers to a set of data which is carried by packets to be sent to the base station. Different packcts may carry the same message when the message is traveling in the network.

Now we analyze and derive the relay probability with which intermediate nodes relaying messages to let them reach the base station with probability $P^*$. $P^*$ is called the success probability.

Suppose there is no packet loss during data transmission.

Case 1: when SourceNode-Gradient=1, which means the source node is one hop from the base station. Once the node broadcasts the message, the base station will receive it immediately. There is no need to relay the message.

Case 2: when SourceNode-Gradient=2, it is a 2-hop message. The source node is two hops from the base station; therefore its 1-hop downstream neighbors need to relay the message.

Suppose the source node has $M_1 (M_1 \geq 1)$ 1-hop nodes as its 1-hop downstream neighbors. These $M_1$ nodes are sibling nodes as the source node is their common 1-hop upstream neighbor. If each of them forwards the message with probability $\eta_1$, the probability that the message reaches the base station is $1 - (1 - \eta_1)^{M_1}$. This probability must satisfy $1 - (1 - \eta_1)^{M_1} \geq P^*$. Hence we have

$$\eta_1 \geq 1 - (1 - P^*)^{1/M_1}. \tag{5.1}$$

For convenience, we set

$$\kappa_1 = 1 - (1 - \eta_1)^{M_1}. \tag{5.2}$$

Eq. (5.1) yields the condition of relay probability for the 1-hop downstream neighbors of the source node. Essentially, the relay probability $\eta_1$ is determined by $M_1$.

Case 3: when SourceNode-Gradient=3, the message's source node is 3 hops from the base station. The message needs to be relayed by intermediate 2-hop nodes and 1-hop nodes.

Suppose the source node has $M_2(M_2 \geq 1)$ 2-hop nodes as its 1-hop downstream neighbors, say $O_1, O_2, \ldots$, and $O_{M_2}$, which also are sibling nodes as they have the same source node as their 1-hop upstream neighbor. Suppose each of them will relay the message with probability $\eta_2$. Moreover, suppose each of these $M_2$ nodes independently has $M_1^{(j)}$ 1-hop downstream neighbors, respectively. Here, $M_1^{(j)} \geq 1$, $j = 1, 2, \cdots, M_2$.

Provided $O_j$ forwards the message, its $M_1^{(j)}$ 1-hop downstream neighbors will receive and relay the message with probability $\eta_1^{(j)}$. $\eta_1^{(j)}$ satisfies Eq. (5.1) for Case 2, that is,

$$\eta_1^{(j)} \geq 1 - (1 - P^*)^{1/M_1^{(j)}}. \tag{5.3}$$

Hence, via $O_j$, the probability that the message successfully reaches the base station is $\kappa_1^{(j)} \eta_2$, where $\kappa_1^{(j)} = 1 - (1 - \eta_1^{(j)})^{M_1^{(j)}}$.

As the source node has $M_2$ 1-hop downstream neighbors, the relay probability $\eta_2$ should satisfy the following condition:

$$\begin{cases} 1 - \displaystyle\prod_{j=1}^{M_2}[1 - \kappa_1^{(j)}\eta_2] \geq P^*, \\ \kappa_1^{(j)} = 1 - (1 - \eta_1^{(j)})^{M_1^{(j)}}. \end{cases} \tag{5.4}$$

As shown in Eq. (5.4), the relay probability $\eta_2$ is dependent on $\{\eta_1^{(j)}\}_{j=1}^{M_2}$. According to Eq. (5.3), $\kappa_1^{(j)} \geq P^*$ holds for $j = 1, 2, \cdots, M_2$. Replacing $\kappa_1^{(j)}$ with $P^*$ in Eq. (5.4), we have

$$1 - \prod_{j=1}^{M_2}(1 - \kappa_1^{(j)}\eta_2) \geq 1 - (1 - P^*\eta_2)^{M_2}.$$

Let $1 - (1 - P^*\eta_2)^{M_2} \geq P^*$. We get another condition for $\eta_2$, that is,

$$\eta_2 \geq \frac{1 - (1 - P^*)^{1/M_2}}{P^*}. \tag{5.5}$$

If $\eta_2$ satisfies Eq. (5.5), then Eq. (5.4) will hold, namely, Eq. (5.5) is more conservative than Eq. (5.4). However, Eq. (5.5) has an obvious advantage as $\eta_2$ is independent on $\kappa_1^{(j)}$ ($j = 1, 2, \cdots, M_2$). In other words, Eq. (5.5) delivers a localized solution for $\eta_2$.

Now we consider the generalized case.

Case 4: when SourceNode-Gradient=$k + 1$. Suppose the source node has $M_k$ 1-hop downstream neighbors, which are $k$-hop nodes and will get involved in relaying the $(k + 1)$-hop message. Denote by $\eta_k$ the relay probability. With the same inductive reasoning and the trick of removing the dependence of $\eta_k$ on $\eta_{k-1}^{(j)}$, we have the condition for $\eta_k$, that is,

$$\eta_k \geq \frac{1 - (1 - P^*)^{1/M_k}}{P^*}. \tag{5.6}$$

To guarantee a success probability $P^*$ for a $(k + 1)$-hop messages, the relay probability of the intermediate $i$-hop nodes, $\eta_i$, is

$$\begin{cases} \eta_i \geq 1 - (1 - P^*)^{1/M_i}, \ i = 1; \\ \eta_i \geq \dfrac{1 - (1 - P^*)^{1/M_i}}{P^*}, \ i = 2, 3, \cdots, k. \end{cases} \tag{5.7}$$

Eq. (5.7) is a localized and conservative solution for $\eta_i$ which depends on $M_i$ only. Such a feature makes the implementation of ProFor much easier.

With Eq. (5.7), we set the relay probability as

$$\eta_i = \begin{cases} 1 - (1 - P^*)^{1/M_i}, \ i = 1; \\ \dfrac{1 - (1 - P^*)^{1/M_i}}{P^*}, \ i = 2, 3, \cdots, k. \end{cases} \tag{5.8}$$

Given $P^*$, the relationship between $\eta_i$ and $M_i$ is shown in Fig. 5.2. A larger $M_i$ means more paths for transmission and results in a smaller $\eta_i$. This observation is consistent with the logic of the real situation. In particular, if there is only one path from the source to the base station which means that $M_1 = 1$ and $M_i = 1 (i = 2, 3, \cdots, k)$, then $\eta_1 = P^*$ and $\eta_i = 1 (i = 2, 3, \cdots, k)$, that is, the 1-hop node relays it with probability $P^*$ and the other intermediate nodes must relay the message. Obviously there exist multiple solutions to make the message reach the base station with probability $P^*$. For instance, let one of the intermediate nodes relay the message with $P^*$ and each of the rest relay it with probability 1. Nevertheless, Eq. (5.8) yields one feasible solution.

$\eta_i$ is determined by $M_i$ and $P^*$. $P^*$ is predefined. Each $(i + 1)$-hop node counts its 1-hop downstream neighbors, which are $i$-hop nodes, at the initialization stage

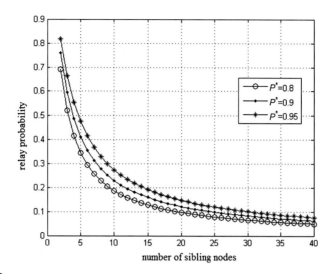

**FIGURE 5.2**

Relay probability.

and then broadcasts its $M_i$ (and $P^*$ if needed) together with the message it relays or initiates so that its 1-hop downstream neighbors can compute $\eta_i$ and decide whether or not to relay the message. The procedure for an intermediate $i$-hop node is described as follows.

**Algorithm 5.1** (Relaying a packet with ProFor protocol).
   Step 1: Receiving one packet.
   Step 2: Checking if this packet being sent from a $(i + 1)$-hop node. If not, going to Step 7.
   Step 3: Taking $P^*$ and $M_i$ from the packet and computing relay probability $\eta_i$ with Eq. (5.8).
   Step 4: Generating a random number $\beta$ which follows uniform distribution $U(0, 1)$.
   Step 5: If $\beta > \eta_i$, going to Step 7.
   Step 6: Updating the packet, i.e. leaving the message unchange but replacing the gradient and the number of 1-hop downstream neighbors with its own. Then relaying the updated packet and going to Step 8.
   Step 7: Discarding the packet and going to Step 8.
   Step 8: Termination.

## 5.1.3 Simulation and analysis

We use network simulator ns2 [10] to simulate ProFor. The simulator ns2 is a discrete event simulator targeted at networking research. It provides substantial support for simulation of TCP, routing, and multicast protocols over wired and wireless networks.

Within an area of $100 \times 100$ square meters, 4000 nodes are randomly and uniformly distributed. The communication radius is 10 meters. The base station is located at the center of the area $(50, 50)$. Fig. 5.3 shows the left lower quarter of the area, in which sensor nodes are marked with different patterns and the same pattern indicates the same gradient. In Fig. 5.3, nodes which are located at the lower left corner are farthest from the base station and their gradient is 10.

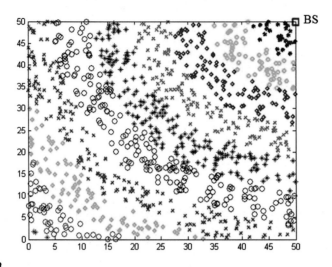

**FIGURE 5.3**

Left lower area of the simulation area (BS: base station).

When a message is being transmitted in the network, a node may receive duplicate packets which carry an identical message. Fig. 5.4 provides an example of this situation, in which both $O_2$ and $O_3$ are 1-hop downstream neighbors of $O_4$; and $O_1$ is 1-hop downstream neighbor of $O_2$ and $O_3$. A packet sent by $O_4$ is relayed by both node $O_2$ and $O_3$. Then $O_1$ will receive two packets separately. According to FroFor, $O_1$ has some probability to relay both of them. Such situations will happen more frequently on nodes which are close to the base station. The FroFor-1 protocol aims to alleviate the traffic load by preventing any node from ever forwarding a message more than once.

To analyze and illustrate the simulation results of ProFor, we compare it with ALL-1. ALL-1 means that a node will rebroadcast any message it receives from its upstream nodes but will not relay an identical message more than once. ALL-1 provides a benchmark to analysis of the performance of ProFor.

We arbitrarily choose one 10-hop node as a source node. Messages are transmitted downstream from the source to the base station. Let the source node send $N_{\text{message}}$ messages in total, here $N_{\text{message}} = 100$. Having fixed the value of $P^*$, we simulate protocols ALL-1, ProFor-1 and ProFor separately, and we calculate the total number of relays and the number of messages received by the base station and compute the total number of relays in the network to estimate the energy consumption. In Fig. 5.5A,

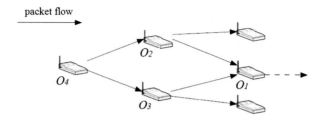

packet flow

**FIGURE 5.4**

Intermediate node receiving duplicate packets.

the success rate is ratio of the number of messages received by the base station to $N_{message}$, and in Fig. 5.5B, the average number of relays is the total number of relays divided by $N_{message}$. The average number of relays is also proportional to the energy consumption on data transmission per message.

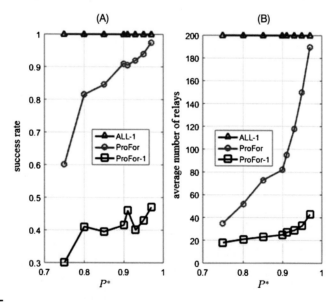

**FIGURE 5.5**

Success rate and average number of relays (gradient of source $=10$).

In Fig. 5.5A and 5.5B, the two lines at the top demonstrate that ALL-1 has a 100% success rate with 200 relays per message; ProFor performs well to secure success probability $P^*$. As $P^*$ is increasing, the success rate with ProFor is even bigger than $P^*$, while the average number of relays also increases. When $P^*$ approaches 100%, the average number of relays increases rapidly and approaches the number incurred by ALL-1. This property indicates that ProFor can save much energy with a small loss of $P^*$. On the other hand, with ProFor-1, the average number of relays has a moderate increase when $P^*$ approaches 1, but the success rate always falls behind $P^*$. Therefore, when $P^* \leq 0.90$, ProFor is a good choice; when $P^*$ is getting larger or approaching 1, we may let a part of the nodes adopt ProFor and the rest adopt ProFor-1 to get a good success rate with less energy consumption.

By simulation experiments, we further explore the relationship between the gradient of the source node and the number of relays per message. Having fixed $P^* = 0.80$, we replicate the above simulations for an arbitrary $i$-hop node, $i = 4, 5, 6, 7, 8, 9$, respectively. We plot the resulting success rate and the average number of relays per message in Fig. 5.6. As the gradient of the source node increases (i.e. the source node is getting farther from the base station), more nodes will be involved in relaying so that the number of relays should increase. Fig. 5.6 shows that ProFor delivers a success rate close to $P^*$ and the average number of relays increases moderately while the performance of ProFor-1 gets worse. Compared with ProFor-1, ProFor provides enough packets for each message to guarantee the success probability.

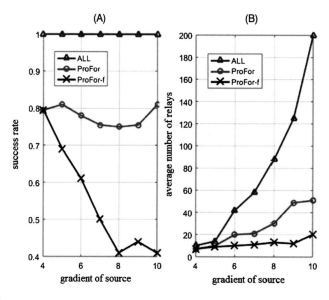

**FIGURE 5.6**

Success rate and average number of relays ($P^* = 0.8$).

## 5.2 Enhanced probabilistic forwarding (EnProFor)

For WSNs, routing protocols should be distributed, have low energy consumption, and be able to cope with frequently changing network topologies [9]. Meanwhile, routing protocols have to tolerate packet loss due to bad radio communication, congestion, packet collision, memory capacity, node failures, etc.

In this section, an enhanced probabilistic forwarding (EnProFor) method is presented for data transmission. Different from ProFor, EnProFor takes into account the packet loss. We derive the conditions for relay probability and explore the asymptotic property of the networks.

### 5.2.1 Analysis of relay probability

In reality, communication with radio frequency is sensitive to many environmental factors. Due to the existence of a gray zone for links between nodes (see [1] for details as regards the communication zone), there is a possibility that a node has some 1-hop downstream neighbors located in its gray zone. If so, the connectivity between the node and its neighbors may become unreliable, so that its neighbor may not receive the packet or receive the packet with error codes. We assume that sensor nodes will simply discard any corrupted packet. For simplicity, the packet loss probability between neighboring nodes is assumed to be $1 - q\,(0 < 1 - q < 1)$.

The network performs an initialization to generate the gradient for each node in the same way as described in Section 5.1.1. Of course, packet loss may also happen on gradient establishment packets. Hence packet loss may let a node get a smaller number of 1-hop downstream neighbors and hence EnProFor could be a little more conservative.

Due to packet loss in the network, we assume the predefined success probability $P^* \leq q$. Now we analyze and derive the sufficient condition for relay probability [6].

Case 1: when SourceNode-Gradient=1, which means the source node is one hop from the base station. When it broadcasts a message, the base station can receive it immediately with probability $q$. Since $P^* \leq q$, no relay is needed.

Case 2: when SourceNode-Gradient=2, the source node is two hops from the base station so that relaying by intermediate 1-hop node(s) is indispensable.

Suppose the source node has $\tilde{M}_1\,(\tilde{M}_1 \geq 1)$ 1-hop downstream neighbors. If each of them relays the message with probability $\tilde{\eta}_1$, then the base station receives it with probability $1 - (1 - \tilde{\eta}_1 q)^{\tilde{M}_1}$. Let $1 - (1 - \tilde{\eta}_1 q)^{\tilde{M}_1} \geq P^*$. Hence

$$\tilde{\eta}_1 \geq \frac{1 - (1 - P^*)^{1/\tilde{M}_1}}{q}. \tag{5.9}$$

Eq. (5.9) stipulates the condition for the relay probability for 1-hop downstream neighbors. $\tilde{\eta}_1$ is dependent on $\tilde{M}_1$ and $q$.

Moreover, we set

$$\tilde{\kappa}_1 = 1 - (1 - \tilde{\eta}_1 q)^{\tilde{M}_1}. \tag{5.10}$$

Case 3: when SourceNode-Gradient=3, the 3-hop message needs to be relayed by 2-hop nodes and 1-hop nodes.

Suppose the source node has $\tilde{M}_2$ 1-hop downstream neighbors, say $O_1, O_2, \cdots$, and $O_{\tilde{M}_2}$, which are 2-hop nodes. Suppose $O_j\ (j = 1, 2, \cdots, \tilde{M}_2)$ will relay the message with probability $\tilde{\eta}_2$ and $O_j$ has $\tilde{M}_1^{(j)}$ 1-hop downstream neighbors, which are

1-hop nodes. Provided $O_j$ sends out the message, its $\tilde{M}_1^{(j)}$ 1-hop downstream neighbors can receive and relay it with probability $\tilde{\eta}_1^{(j)}$. $\tilde{\eta}_1^{(j)}$ is defined by Eq. (5.9), namely

$$\tilde{\eta}_1^{(j)} \geq \frac{1 - (1 - P^*)^{1/\tilde{M}_1^{(j)}}}{q}.$$

Then the probability for the message successful arriving at the base station via $O_j$ is $\tilde{\kappa}_1^{(j)} \tilde{\eta}_2$, where $\tilde{\kappa}_1^{(j)} = 1 - (1 - \tilde{\eta}_1^{(j)} q)^{\tilde{M}_1^{(j)}}$.

The probability that the message fails to arrive at the base station is $\prod\limits_{j=1}^{\tilde{M}_2}(1 - \tilde{\kappa}_1^{(j)} \tilde{\eta}_2 q)$. Hence the success probability is $1 - \prod\limits_{j=1}^{\tilde{M}_2}(1 - \tilde{\kappa}_1^{(j)} \tilde{\eta}_2 q)$. $\eta_2$ should satisfy

$$\begin{cases} 1 - \prod\limits_{j=1}^{\tilde{M}_2}(1 - \tilde{\kappa}_1^{(j)} \tilde{\eta}_2 q) \geq P^*, \\ \tilde{\kappa}_1^{(j)} = 1 - (1 - \tilde{\eta}_1^{(j)} q^2)^{\tilde{M}_1^{(j)}}. \end{cases} \tag{5.11}$$

With (5.10), $\tilde{\kappa}_1^{(j)} = 1 - (1 - \tilde{\eta}_1^{(j)} q)^{\tilde{M}_1^{(j)}} \geq P^*$ holds for $j = 1, 2, \cdots, \tilde{M}_2$. In Eq. (5.11), by replacing $\tilde{\kappa}_1^{(j)}$ with $P^*$, we have

$$1 - \prod\limits_{j=1}^{\tilde{M}_2}(1 - \tilde{\kappa}_1^{(j)} \tilde{\eta}_2 q) \geq 1 - (1 - P^* \tilde{\eta}_2 q)^{\tilde{M}_2}.$$

Let $1 - (1 - P^* \tilde{\eta}_2 q)^{\tilde{M}_2} \geq P^*$, hence we have the condition for $\tilde{\eta}_2$ as

$$\tilde{\eta}_2 \geq \frac{1 - (1 - P^*)^{1/\tilde{M}_2}}{P^* q}. \tag{5.12}$$

Eq. (5.12) is not dependent on $\{\tilde{\eta}_1^{(j)}, j = 1, 2, \cdots, \tilde{M}_2\}$. It provides a localized solution for $\tilde{\eta}_2$.

Case 4: when SourceNode-Gradient $= k + 1$ in which a $(k + 1)$-hop message relaying by $\tilde{M}_k(\tilde{M}_k \geq 1)$ $k$-hop nodes. Each of them relays the message with probability $\tilde{\eta}_k$. Similar to the analysis in Case 3, the condition for $\tilde{\eta}_k$ is

$$\tilde{\eta}_k \geq \frac{1 - (1 - P^*)^{1/\tilde{M}_k}}{P^* q}. \tag{5.13}$$

In general, to let a $(k+1)$-hop message successfully arrive at the base station with probability $P^*$, the condition for the relay probability $\tilde{\eta}_i$ is

$$
\begin{cases}
\tilde{\eta}_i \geq \dfrac{1-(1-P^*)^{1/\tilde{M}_i}}{q}, & i = 1; \\[3mm]
\tilde{\eta}_i \geq \dfrac{1-(1-P^*)^{1/\tilde{M}_i}}{P^*q}, & i = 2, 3, \cdots, k.
\end{cases}
\tag{5.14}
$$

This probability determined by $\tilde{M}_i$, success rate $P^*$ and $q$. We can set

$$
\tilde{\eta}_i =
\begin{cases}
\dfrac{1-(1-P^*)^{1/\tilde{M}_i}}{q}, & i = 1; \\[3mm]
\dfrac{1-(1-P^*)^{1/\tilde{M}_i}}{P^*q}, & i = 2, 3, \cdots, k.
\end{cases}
\tag{5.15}
$$

This probabilistic forwarding protocol takes into account the possibility of packet loss. Hence it is an enhanced version of ProFor. With EnProFor, the procedure for an intermediate $i$-hop node is described as follows.

**Algorithm 5.2** (Relaying a packet with EnProFor protocol ($q$ and $P^*$ are given)).

Step 1: Receiving one packet.

Step 2: Checking if this packet has been sent from a $(i+1)$-hop node. If not, going to Step 7.

Step 3: Taking $P^*$ and $M_i$ from the packet and computing relay probability $\eta_i$ with Eq. (5.15).

Step 4: Generating a random number $\beta$ which follows the uniform distribution $U(0, 1)$.

Step 5: If $\beta > \eta_i$, going to Step 7.

Step 6: Updating the packet, i.e. leaving the message unchange but replacing the gradient and the number of 1-hop downstream neighbors with its own. Then relaying the updated packet and going to Step 8.

Step 7: Discarding the packet and going to Step 8.

Step 8: Termination.

## 5.2.2 Simulation and analysis

We use the same settings and follow the same way as described in Section 5.1.3 to carry out simulations for EnProFor. Similar to ProFor-1, EnProFor-1 means that if a relay node receives multiple packets of one message, it will forward the message at most once.

Figs. 5.7A and 5.7B show that, when $q < 1$, ProFor fails to achieve the predefined $P^*$ but EnProFor can make it. Of course, EnProFor needs more relays. Compared to

EnProFor, EnForFor-1 needs less relays but its success rate is smaller. As $P^*$ gets larger, the success rate lags much further behind.

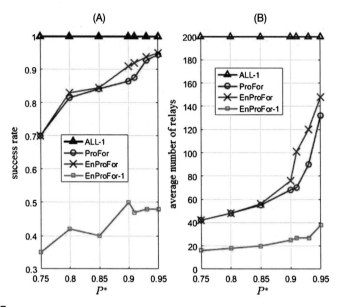

**FIGURE 5.7**

Success rate and average number of relays (gradient of source $=10$ and $q = 0.95$).

Figs. 5.8A and 5.8B show that EnProFor performs very well and meets the requirement. As the gradient of the source node increases, there are more nodes getting involved in relaying the message. The average number of relays required by ALL-1 increases sharply, but the number required by EnProFor increases mildly. Hence, for a large scale WSN, EnProFor is much more energy efficient.

Since the success rate by EnProFor-1 is much smaller than $P^*$, we introduce EnProFor-$n$, which lets a node forward packets of one message at most $n$ times.

By simulation, we compare EnProFor, EnProFor-1, EnProFor-2 and EnProFor-3. As shown in Figs. 5.9A and 5.9B, with a larger $n$, the success rate obtained with EnProFor-$n$ is getting closer to the rate with EnProFor. And the corresponding number of relays also approaches the number required by EnProFor. However, on average, EnProFor-3 needs less relays than EnProFor does, especially when $P^*$ goes to 1.

Figs. 5.10A and 5.10B show the same features of EProFor-$n$ as discussed above and the growing gap of success rate and average number of relays between EnProFor and EnProFor-1, 2, 3 as the gradient of source node increases.

The results of the simulation indicate that EnProFor achieves the predefined success rate $P^*$ when considering packet loss. The success rate obtained with EnProFor-3 is almost the same as the rate with EnProFor but EnProFor-3 needs less relays.

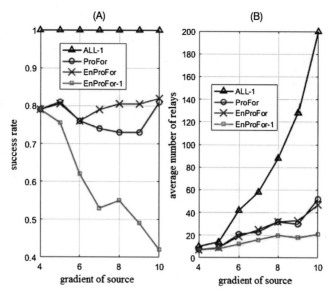

**FIGURE 5.8**

Success rate and average number of relays ($P^* = 0.8$ and $q = 0.95$).

## 5.3 Analysis of energy consumption

In this section, we analyze the relay procedure to estimate the number of relays.

### Relaying by sibling nodes

Consider the procedure of relaying message. The number of $i$-hop sibling nodes which receive the packet sent by their common 1-hop upstream neighbor is $M_i$. Each of these $M_i$ sibling nodes computes the relay probability $\eta_i$ and then generates a random number following the uniform distribution $U(0, 1)$; if the number falls in $(0, \eta_i]$, then the node will relay the packet. Suppose there are $m_i$ of these $M_i$ sibling nodes that eventually relay the packet and the rest, $M_i - m_i$ nodes, discard the packet. $m_i$ is a random variable with the mean of

$$
\mathrm{E}(m_i) = M_i \eta_i = \begin{cases} \dfrac{M_i(1 - (1 - P^*)^{1/M_i})}{q}, & i = 1; \\[3ex] \min\left\{ \dfrac{M_i(1 - (1 - P^*)^{1/M_i})}{P^*q}, M_i \right\}, & i \geq 2, \end{cases} \tag{5.16}
$$

where operator E is for taking the expectation. We compute $\mathrm{E}(m_i)$ with Eq. (5.16) and plot it in Fig. 5.11, which shows that $\mathrm{E}(m_i)$ increases monotonically with $M_i$; and when $M_i$ goes to infinity, $\mathrm{E}(m_i)$ will converge to a finite number instead of growing unlimitedly. In other words, there exists a finite upper bound for $\mathrm{E}(m_i)$. In fact, when

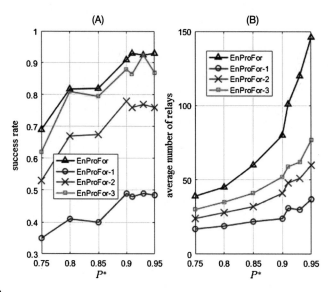

**FIGURE 5.9**

Success rate and average number of relays (gradient of source $=10$ and $q=0.95$).

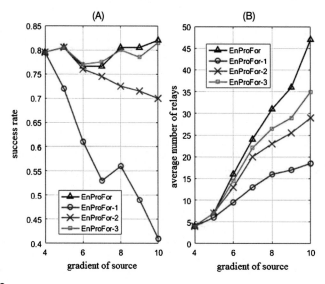

**FIGURE 5.10**

Success rate and average number of relays ($P^*=0.8$ and $q=0.95$).

$M_i \rightarrow \infty$, with the Taylor series we have

$$(1-P^*)^{1/M_i} = 1 + \frac{\ln(1-P^*)}{1!M_i} + \frac{[\ln(1-P^*)]^2}{2!M_i^2} + o\left(\frac{1}{M_i^2}\right),$$

$$1 - (1 - P^*)^{1/M_i} = -\frac{\ln(1 - P^*)}{1!M_i} - \frac{[\ln(1 - P^*)]^2}{2!M_i^2} + o\left(\frac{1}{M_i^2}\right). \qquad (5.17)$$

If we denote by $\overline{E(m_i)}$ the upper bound of $E(m_i)$, with Eq. (5.17) and Eq. (5.16), by taking the limit we obtain

$$\overline{E(m_i)} = \lim_{M_i \to \infty} E(m_i) = \begin{cases} \dfrac{-\ln(1 - P^*)}{q}, & i = 1; \\ \min\left\{\dfrac{-\ln(1 - P^*)}{P^*q}, M_i\right\}, & i \geq 2. \end{cases} \qquad (5.18)$$

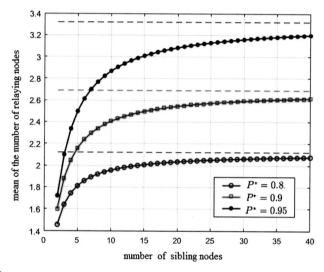

**FIGURE 5.11**

Mean of the number of relay nodes ($q = 0.95$).

If $q = 0.95$, when $P^* = 0.80, 0.90, 0.95$, the corresponding $\overline{E(m_i)}$ is 2.12, 2.69, 3.32, respectively. They are shown with three horizontal broken lines in Fig. 5.11. As $M_i$ is getting large, $m_i$ is approaching its upper bound $\overline{E(m_i)}$.

## Simulation and analysis of the number of relays

We use the same settings and follow the same way as described in Section 5.1.3 to carry out simulations for EnProFor.

Let $q = 0.95$ and $P^* = 0.9$. We simulate the EnProFor protocol and collect statistics about the relay procedure of intermediate nodes and list them in Table 5.1. For each S-node (abbr. of source node), we count the number of its 1-hop downstream neighbors, $M_i$, the number of sibling nodes. Each S-node broadcasts a number of packets (in the third column), we calculate the total number of relays by its 1-hop downstream neighbors (in the fourth column) and the average number of relays per packet sent by S-node (in the fifth column).

**Table 5.1** Partial statistics for source nodes ($q = 0.95$, $P^* = 0.9$).

| S-node ID | $M_i$ | Num. of packets | Num. of relays | Average num. of relays |
|---|---|---|---|---|
| 43 | 17 | 10 | 24 | 2.40 |
| 60 | 17 | 2 | 2 | 1.00 |
| **131** | **23** | **8** | **24** | **3.00** |
| 152 | 2 | 151 | 171 | 1.13 |
| 156 | 4 | 20 | 26 | 1.30 |
| 190 | 2 | 13 | 12 | 0.92 |
| 200 | 18 | 1 | 3 | 3.00 |
| 201 | 1 | 18 | 9 | 0.50 |
| 213 | 1 | 126 | 98 | 0.78 |
| 235 | 5 | 13 | 20 | 1.54 |
| 261 | 7 | 5 | 9 | 1.80 |
| 317 | 7 | 11 | 23 | 2.09 |
| 352 | 17 | 11 | 16 | 1.45 |
| 357 | 14 | 1 | 3 | 3.00 |
| 365 | 8 | 14 | 15 | 1.07 |
| 377 | 3 | 4 | 6 | 1.50 |
| 389 | 2 | 48 | 56 | 1.17 |
| **424** | **12** | **7** | **21** | **3.00** |
| 461 | 2 | 185 | 231 | 1.25 |
| 493 | 29 | 3 | 1 | 0.33 |

We observe that, no matter how big the value of $M_i$ is, the average number of relays is less than 3.32. This agrees with our analysis as regards $\overline{E(m_i)}$. For example, S-node 131 has 23 downstream neighbors while S-node 424 has 12 downstream neighbors. The former is nearly twice as much as the latter. S-node 131 broadcasts 8 packets while S-node 424 broadcasts 7 packets but the average number of relays is 3 for both of them.

As one message is relayed to 1-hop downstream nodes, the number of packets of the message increases at most $\overline{E(m_i)}$ times, which is less than $M_i$. Hence, for a large scale WSN in which sensor nodes are densely deployed, EnProFor costs much less energy than ALL-1 does (see Section 5.2.2).

For a relay node, its relay probability is determined by Eq. (5.15). Fig. 5.12 shows that when there are fewer sibling nodes the relay probability will increase; when $P^*$ is fixed, a larger $q$ leads to a smaller relay probability. The relay probability is sensitive to $M_i$ when $M_i$ is small and it converges when $M_i$ is getting larger.

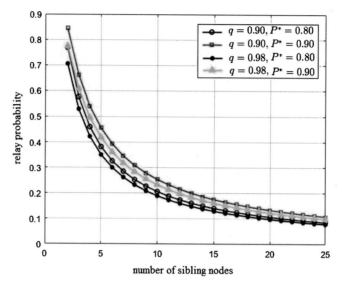

**FIGURE 5.12**

Relay probability.

Given $q = 0.95$ and $P^* = 0.9$, we simulate the EnProFor protocol and collect statistics of the relay procedure and list a part of the data in Table 5.2, in which the R-node ID is the ID of the relay node. The R-node follows Algorithm 5.2. The second column records the gradient of the R-node; the third column is the number of packets received by the R-node from its 1-hop upstream neighbor. The fourth column records the total number of packets that the R-node has relayed. The last column is the average number of the R-node's sibling nodes, i.e. the average of $M_i$ encoded in packets received by the R-node.

Nodes which are close to the base station usually have to relay more packets and hence spend more energy. For example, the gradient of R-node 1112 is 2; it receives a large number of packets but it has less sibling nodes. With Eq. (5.15), we know that its relay probability is 0.5638 as $q = 0.95$, $P^* = 0.9$ and $M_i = 3.5$. In the simulation, it relays 253 of 493 packets, the relay rate is 51.31%.

Nodes with more sibling nodes are usually less active in relaying as they share relay tasks with their sibling nodes. For example, the relay rate for R-node 1171 is smaller than that for R-node 1271 and R-node 1018.

Even nodes are in idle mode, they also consume much energy (e.g. see Table 1 of the radio power characterization in Ref. [13]). Hence redundant nodes should be turned OFF to save energy [2,12,13].

According to the earlier analysis, Eq. (5.18) provides the theoretical upper bound of $E(m_i)$, $\overline{E(m_i)}$. Hence the rest, $M_i - \overline{E(m_i)}$ sibling nodes, are redundant. To balance energy consumption among the sibling nodes, these $M_i$ nodes should take turns to relay messages.

**Table 5.2** Partial statistics for relay nodes ($q = 0.95$, $P^* = 0.9$).

| R-node ID | gradient | Num. of packets received | Num. of relays | Relay rate | Average of $M_i$ |
|---|---|---|---|---|---|
| 128 | 7 | 2 | 1 | 50% | 36 |
| **1271** | **7** | **87** | **21** | **33.33%** | **14.14** |
| 1255 | 4 | 2 | 1 | 50% | 9 |
| 1254 | 4 | 95 | 30 | 31.57% | 6.25 |
| 1241 | 5 | 93 | 19 | 20.43% | 14.5 |
| 124 | 1 | 11 | 4 | 36.36% | 4.67 |
| 1220 | 6 | 25 | 7 | 28% | 6 |
| 1189 | 6 | 61 | 16 | 26.22% | 6.17 |
| **1171** | **7** | **21** | **4** | **19.05%** | **20** |
| 1163 | 6 | 29 | 9 | 31.03% | 6.75 |
| 1114 | 6 | 5 | 2 | 40% | 4 |
| **1112** | **2** | **493** | **253** | **51.31%** | **3.5** |
| 1066 | 5 | 127 | 42 | 33.07% | 5.25 |
| 1034 | 4 | 25 | 8 | 32% | 6 |
| **1018** | **7** | **93** | **22** | **23.65%** | **8.57** |

The state switching scheme presented in Section 4.1 may be applied with $\rho = \lceil \overline{E(m_i)} \rceil / M_i$ as the working probability, where $\lceil x \rceil$ represents the smallest integer which is equal to or bigger than $x$. Accordingly, an intermediate node should use $\lceil \overline{E(m_i)} \rceil$ instead of $M_i$ to determine its relay probability or simply relay any packet received from its 1-hop upstream neighbors.

## Multiple base stations

In a WSN, nodes close to the base station have to spend much more energy on relaying messages. Once their energy is used up, the network cannot work any more. Such a situation generally exists in various network topologies [4]. Deploying more nodes around the base station should be helpful. However, it may be impossible or too expensive to deploy more nodes in a particular region. As discussed in previous chapters, a randomly and uniformly deployed WSN usually has a good coverage performance. Moreover, one static base station may be unable to efficiently collect data for the long term. An intuitive solution is configuring multiple base stations such that nodes can send data to the nearest one, or base stations take turns working to balance the energy consumption among all nodes.

When implementing ProFor or EnProFor with multiple base stations, every node's gradient should be initialized with respect to each base station separately. Hence every node knows the corresponding 1-hop upstream neighbor and the number of 1-hop downstream neighbors.

## Message priority

Messages generated by different nodes or with different contents may be of higher or lower importance. ProFor and EnProFor can distinguish the rank of importance with different success probabilities $P^*$. A bigger $P^*$ implies a higher importance.

## 5.4 Summary

ProFor and EnProFor protocols are distributed and localized protocols. They can adapt to a topological change of the network and work robustly against individual node failure. Of course, probabilistic forwarding inherently runs the risk of losing a message. However, both analysis and simulations show that ProFor and EnProFor work very well to guarantee a high success probability for data transmission.

In practical situations, ProFor and EnProFor are much easier and better for implementation. The results in this chapter are widely applicable to different settings such as multiple base stations and message priority. The basic idea and framework of ProFor or EnProFor are also applicable to heterogeneous WSNs.

The theoretical analysis and simulation demonstration not only reveal the underlying features but also provide deep insights of a large scale WSN. They have great potential to improve the scheduling at a medium access control layer to gain better energy-efficiency.

## References

[1] G. Anastasi, E. Borgia, M. Conti, et al., Understanding the real behavior of Mote and 802.11 ad hoc networks: an experimental approach, Pervasive and Mobile Computing 1 (2005) 237–256.

[2] ASH Transceiver Designer's Guide http://www.rfm.com, May 2004.

[3] L.C. Barrett, S.J. Eidenbenz, L. Kroc, M. Marathe, J.P. Smith, Parametric probabilistic sensor network routing, in: Proceedings of the 2nd ACM International Conference on Wireless Sensor Networks and Applications, ACM, NY, 2003.

[4] Z. Cai, H. Zhang, Research on node deploying scheme in layered wireless sensor networks (in Chinese), Computer Engineering and Applications 44 (35) (2008) 112–114.

[5] X. Chen, Y.C. Ho, J. Zhang, Probabilistic forwarding (ProFor) for large scale sensor networks, IEEE International Conference on Networking (2006) 963–968.

[6] X. Chen, X. Wang, An enhanced probabilistic scheme for data transmission in large scale sensor networks, Frontiers of Electrical and Electronic Engineering in China 6 (3) (2011) 481–485.

[7] Z. Haas, J.Y. Halpern, L. Li, Gossip-based ad hoc routing, in: Proceedings of IEEE INFOCOM, New York, 2002.

[8] W. Heinzelman, A. Chandrakasan, H. Balakrishnan, Energy-efficient communication protocol for wireless sensor networks, in: Proceeding of the Hawaii International Conference on System Sciences, Hawaii, January 2000.

[9] H. Karl, A. Willig, Protocols and Architectures for Wireless Sensor Network, 1st edition, John Wiley & Sons, Ltd, London, 2005.

[10] http://www.isi.edu/nsnam/ns/.

[11] R. Nagpal, H. Shrobe, J. Bachrach, Organizing a Global Coordinate System From Local Information on an Ad Hoc Sensor Network, Lecture Notes in Computer Science, vol. 2634, Springer, Berlin, 2003, pp. 553–569.

[12] V. Raghunathan, C. Schurgers, S. Park, M. Srivastava, Energy-aware wireless sensor networks, IEEE Signal Processing 19 (2) (March 2002) 40–50.

[13] C. Schurgers, V. Tsiatsis, S. Ganeriwal, et al., Optimizing sensor networks in the energy–latency–density design space, IEEE Transactions on Mobile Computing 1 (1) (2002) 70–80.

[14] J. Wu, L. Chen, P. Yan, J. Zhou, H. Jiang, A new reliable routing method based on probabilistic forwarding in wireless sensor network, in: Proceedings of the Fifth International Conference on Computer and Information Technology, CIT'05, CIT, Shanghai, 2005.

# Stochastic scheduling algorithms

## CONTENTS

For a WSN in which sensor nodes are randomly deployed in the area of interest, its performance is essentially determined by the number of working nodes. Since each node has limited energy, some nodes in the network may run out of their energy earlier than others. Maintaining the number of working nodes at the same level will result in a stable network performance.

In Chapter 3, two coverage protocols, percentage coverage configuration protocol [1] and standing guard protocol [2], are discussed. Both of them adopt a way of working in rounds as follows. Suppose $T$ is the period length. All nodes are in a working state at the beginning of a period. In the coverage configuration phase, each node decides to work or sleep in the remaining time of the period by negotiation with its neighbors. At the end of the period, sleeping nodes will switch to working state and all nodes enter into the coverage configuration phase of the 2nd period to decide their state. Each node repeats this procedure for the 3rd period, the 4th period, and so on till its energy is exhausted, as shown in Fig. 6.1.

However, the state switching scheme presented in Section 4.1 is much easier for implementation and more energy-efficient. It does not need time synchronization and each node operates independently. In this chapter, starting from the state switching scheme, we investigate the procedure of energy consumption for each sensor node and propose an algorithm for the working probability to guarantee that the number of working nodes keeps at the same level in each period [3].

**Randomly Deployed Wireless Sensor Networks.** https://doi.org/10.1016/B978-0-12-819624-3.00011-2

**FIGURE 6.1**

Working in rounds mode.

## 6.1 Model description

First of all, the following assumptions are imposed on the network and sensor nodes.

**Assumption 6.1.** *The area of interest is a unit square.*

**Assumption 6.2.** *Sensor nodes are randomly and uniformly distributed in the area of interest.*

**Assumption 6.3.** *Sensor nodes are homogeneous.*

**Assumption 6.4.** *A boolean sensing model is adopted.*

Suppose the network has enough sensor nodes to achieve the desired performance, for instance, to have 90% of the area of interest being 2-covered.

To save energy, each sensor node independently switches its state, with probabilities $\rho$ for working and $1 - \rho$ for sleeping, respectively, and keeps its state at least for a period before taking the chance to change its state. Sensor nodes do not need time synchronization (also see Section 4.1).

In real applications, the area of interest may take various shapes. Assumption 6.1 is for convenience of analysis. We make it without loss of generality. The results based on the assumption can be easily applied to other shapes.

**Definition 6.1.** *If a node has not used up its energy and can work normally, it is called an effective node. On the contrary, if a node has exhausted its energy or becomes failure, it is called an ineffective node.*

The network lifetime may be measured with the period length. Namely, it is the maximal number of periods in which the total number of effective nodes is always bigger than or equal to the number of working nodes required by the network performance.

### Upper bound of the lifetime

Suppose a network consists of $N$ nodes. The initial energy is $E_o$ for each node. Hence the total energy the network has is $NE_o$. Denote $Q_{on}$ and $Q_{off}$ as the power consumption per time unit when a node is working and sleeping, respectively. Then the maximum working time length for each node is $E_o/Q_{on}$.

Suppose that the period length is $T$ and the network needs $M_0$ working nodes in each period. Then the energy consumption per period for the network is $M_0 Q_{on} T$.

Usually, we assume $Q_{off} << Q_{on}$ (see Table 1 of the radio power characterization in [4]). One upper bound of the network lifetime is

$$\hat{L}_{sn} = \left( \frac{NE_o}{M_0 Q_{on} T} \right) T = \frac{NE_o}{M_0 Q_{on}}, \tag{6.1}$$

as no energy consumption on state switching is taken into account.

$M_0$ is crucial to prolong the network lifetime. However, $M_0$ hinges on the quality of service of a WSN. In the case that the point coverage is of interest, according to the analysis in Section 2.2.1, with Eq. (2.1), the point $k$-coverage probability $\alpha_k$ is

$$\alpha_k = 1 - \sum_{i=0}^{k-1} \binom{M_0}{i} \beta^i (1 - \beta)^{M_0 - i}, \tag{6.2}$$

where $\beta$ is the ratio of node's sensing area to the area of interest. Given $k$ and $\alpha_k$, we know $M_0$. If the network is used to detect targets, some formulas for the detection probability presented in Chapter 4 may help.

## 6.2 Stochastic scheduling

For the stochastic scheduling model, we first analyze the energy consumption and the network lifetime, then we develop an algorithm to determine the dynamic working probability.

### 6.2.1 Analysis of energy consumption

According to the state switching scheme, every sensor node decides its state at the beginning of each period with a corresponding probability and then decides on working or sleeping within the current period. We assume that state switching either from working to sleeping or from sleeping to working spends time $t_s$ and needs extra energy $e_s$.

For each sensor node, its energy consumption in the current period depends not only on its current state, denoted by $S_c$, but also on its state in the previous period, denoted by $S_1$. Again, with $Q_{on}$ and $Q_{off}$ denoting the power consumption per time unit when working and sleeping, respectively, we discuss energy consumption according to the following four possible cases.

Case 1: when $S_1 = S_c =$ working, no state switching is needed for the node and its energy consumption in the current period is

$$E_{ww} = Q_{on} T.$$

Case 2: when $S_1 =$ sleeping and $S_c =$ working, the sensor node switches state once and its energy consumption in the current period is

$$E_{sw} = Q_{on} T + e_s.$$

Case 3: when $S_l$ = working and $S_c$ = sleeping, the sensor node switches state once and its energy consumption in the current period is

$$E_{ws} = Q_{off}(T - t_s) + Q_{on}t_s + e_s.$$

Case 4: when $S_l$ = sleeping and $S_c$ = sleeping, the sensor node switches state twice and its energy consumption in the current period is

$$E_{ss} = Q_{off}(T - t_s) + Q_{on}t_s + 2e_s.$$

Obviously, $E_{sw} > E_{ww}$ and $E_{ss} > E_{ws}$. For each period, the maximal energy consumption is $E_{sw}$ and the least one is $E_{ws}$. If we ignore the energy cost on state switching, namely $e_s = 0$, then we have $E_{sw} = E_{ww}$ and $E_{ss} = E_{ws}$.

If a sensor node works in the previous period and the current period with probability $\rho_l$ and $\rho_c$, respectively, then the average energy it consumed in the current period is

$$
\begin{aligned}
E_c &= \rho_l \rho_c E_{ww} + (1 - \rho_l)\rho_c E_{sw} + \rho_l(1 - \rho_c)E_{ws} + (1 - \rho_l)(1 - \rho_c)E_{ss} \\
&= \rho_c Q_{on}T + (1 - \rho_c)Q_{off}T + (1 - \rho_c)(Q_{on} - Q_{off})t_s + (2 - \rho_l - \rho_c)e_s.
\end{aligned}
$$

Hence the average energy it consumes per time unit is

$$\bar{Q} = E_c/T = \rho_c Q_{on} + (1 - \rho_c)Q_{off} + (1 - \rho_c)(Q_{on} - Q_{off})\frac{t_s}{T} + (2 - \rho_l - \rho_c)\frac{e_s}{T}.$$

$$(6.3)$$

Both $t_s$ and $e_s$, are mainly determined by sensor node's physical property and independent of $T$. Hence, by either reducing the working probability $\rho_c$ or extending the period length $T$, the average energy consumption will reduce to get a longer network lifetime.

## 6.2.2 Dynamic tuning of working probability

In a randomly and uniformly deployed WSN, its performance is largely determined by the number of working nodes. Hence the requirement on the network performance will be eventually transformed to a requirement on the number of working nodes.

For example, given the value of the point $k$-coverage probability, $P^*$, being the requirement on coverage, the number of working nodes, $M_0$, should satisfy the condition defined with Eq. (6.2), that is,

$$P^* = 1 - \sum_{i=0}^{k-1} \binom{M_0}{i} \beta^i (1 - \beta)^{M_0 - i}.$$

Generally, denote by $N_i$ ($i = 1, 2, \cdots$) the number of effective nodes at the beginning of the $i$th period. Then we have the working probability in the 1st period for each sensor node,

$$\rho_1 = \frac{M_0}{N_1} = \frac{M_0}{N}. \tag{6.4}$$

For the working probability $\rho_2$ in the 2nd period, we first compute the total number of effective nodes $N_2$ and then get

$$\rho_2 = \frac{M_0}{N_2}. \tag{6.5}$$

However, due to the randomness of node scheduling, it is hard to get the exact value of $N_2$. We use the expectation of $N_2$, $E(N_2)$, to replace $N_2$. Therefore

$$\rho_2 \approx \frac{M_0}{E(N_2)}. \tag{6.6}$$

In general, denote by $\phi_m$ the probability that a node still has energy at the end of the $m$th period. Since each node independently decides its state (working or sleeping), at the beginning of the 2nd period, the average number of effective nodes is

$$E(N_2) = N_1\phi_1 = N\phi_1; \tag{6.7}$$

combining Eq. (6.7) with Eq. (6.6), we have the working probability for the 2nd period

$$\rho_2 \approx \frac{M_0}{N\phi_1}. \tag{6.8}$$

In the 1st period, each node has probability $\rho_1$ to work and $1 - \rho_1$ to sleep. According to the analysis in the previous section, the maximal energy consumption in the 1st period is $Q_{on}T = E_{ww}$. Hence, at the end of the 1st period, the remaining energy for each sensor node is at least $E_o - E_{ww}$. If $E_o - E_{ww} > 0$, then the node is effective at the beginning of the 2nd period, namely $\phi_1 = 1$; otherwise, if $E_o - E_{ww} \leq 0$, then the node which works in the 1st period will use up its energy to become ineffective, hence the probability that the node is effective at the beginning of the 2nd period is $\phi_1 = 1 - \rho_1$. We get the working probability for the 2nd period, $\rho_2$, with Eq. (6.8).

If a node has been in sleeping state for each period, its maximal lifetime in terms of the number of periods is $L_{max} = \lfloor E_o/E_{ss} \rfloor$; in contrast, if a node has been in the working state for each period, it will exhaust its energy soon and its minimal lifetime in terms of the number of periods is $L_{min} = \lceil E_o/E_{ww} \rceil$. When the number of periods, $m$, falls in $(1, L_{min})$, all nodes in the network are effective; when $m \in [L_{min}, L_{max}]$, a part of nodes in the network will become ineffective and the number of effective nodes decreases gradually; when $m > L_{max}$, all nodes will die.

Suppose that, in the first $m(0 < m \leq L_{max})$ periods, the working probabilities are $\rho_1, \rho_2, \cdots, \rho_m$, respectively. At the end of the $m$th period, for each node, its maximal energy consumption is $mE_{ww}$. If $mE_{ww} < E_o$, every node in the network is still effective, that is, $\phi_m = 1$. If $m E_{ww} \geq E_o$ and a node is still effective, then this node does not work in some of these $m$ periods, otherwise it has exhausted its energy already.

Denote by $u_m$ the number of periods in which an effective node is in the working state, then the maximum of $u_m$, $\max\{u_m\}$, satisfies

$$\max\{u_m\} = \{u | u \leq m, u E_{ww} + (m-u) E_{ws}$$
$$\leq E_o, E_o < (u+1) E_{ww} + (m-u-1) E_{ws}\}. \qquad (6.9)$$

Denote by $v_i$ the state of a node in its $i$th period, i.e. $v_i = 1$ represents working and $v_i = 0$ means sleeping. Therefore, the probability that the node is in the working or sleeping state is

$$q(v_i) = \begin{cases} \rho_i, & v_i = 1; \\ 1 - \rho_i, & v_i = 0. \end{cases} \qquad (6.10)$$

Hence the probability that the node is effective at the end of $m$th period, $\phi_m$, is

$$\phi_m = \sum_{\sum_{i=1}^{m} v_i \leq \max\{u_m\}} \prod_{i=1}^{m} q(v_i). \qquad (6.11)$$

So the expected number of effective nodes at the beginning of the $(m+1)$th period, $E(N_{m+1})$, is

$$E(N_{m+1}) = N_m \phi_m. \qquad (6.12)$$

To guarantee that there are about the same number of working nodes in each period in the network, the working probability for the $(m+1)$th period, $\rho_{m+1}$, has to be

$$\rho_{m+1} = \frac{M_0}{E(N_{m+1})}. \qquad (6.13)$$

The sequence of working probability $\rho_i, i = 1, 2, \cdots$, is non-decreasing. Especially in later stages, $\rho_i$ will increase as the number of effective nodes decreases. In summary, the scheduling algorithm for dynamic working probability is as follows.

**Algorithm 6.1** (Computing the dynamic working probability).

Step 1: Computing the number of working nodes according to the requirement on the network performance (e.g. Eq. (6.2)); determining the initial working probability with Eq. (6.4); setting $m = 1$.

Step 2: With Eq. (6.11) and Eq. (6.12), computing the expected number of effective nodes at the end of $m$th period; with Eq. (6.13), calculating the working probability for the $(m+1)$th period; $m = m + 1$.

Step 3: Repeating Step 2.

Algorithm 6.1 is easy to implement. Sensor nodes do not need to mutually exchange information. Hence the algorithm can save energy on communication.

### 6.2.3 Performance test

By simulation experiments, we analyze and compare the impacts of a constant working probability and dynamic working probability on the coverage performance. We will focus on the number of effective nodes, the working probability and the number of working nodes.

Parameters used in simulation experiments are listed in Table 6.1. As $M_0 = 200$, the network needs 200 working nodes. With Eq. (6.4), the initial working probability for each node is $\rho_1 = 200/1000 = 0.2$. Moreover, with Eq. (6.12) and Eq. (6.13), the expected number of effective nodes at the beginning of each period and the working probability in each period are shown in Fig. 6.2 and Fig. 6.3, respectively.

**Table 6.1**  Parameters in simulation.

| Notation | Meaning | Value |
|:---:|:---:|:---:|
| $N$ | the total number of nodes | 1000 |
| $E_o$ | total energy of a node | 1500 |
| $Q_{off}$ | power consumption per time unit for each sleeping node | 0.001 |
| $Q_{on}$ | power consumption per time unit for each working node | 0.05 |
| $e_s$ | energy consumption for state switching once | 0.005 |
| $T$ | the period length | 5000 |
| $t_s$ | time length of configuration phase | 10 |
| $M_0$ | number of working nodes required for the network performance | 200 |

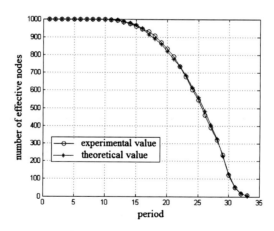

**FIGURE 6.2**

Number of effective nodes in each period.

Fig. 6.2 shows the number of effective nodes at the beginning of each period in the network, where the theoretical value is the expected number computed with Eq. (6.12). As time goes on, more and more nodes exhaust their energy and the number of effective nodes decreases, the theoretical value agrees well with the experimental result.

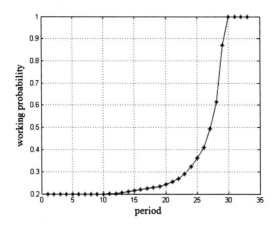

**FIGURE 6.3**

Working probability for each period.

Eq. (6.13) delivers the working probability for each period. As shown in Fig. 6.3, the working probability is increasing to compensate for the possible performance degradation caused by an ineffective node.

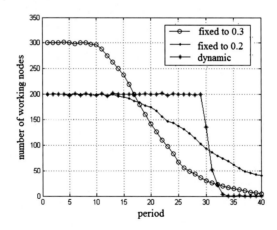

**FIGURE 6.4**

Number of working nodes in each period.

Due to the requirement on the network performance, the working probability cannot be smaller than 0.2. With the working probability fixed to 0.2 and 0.3, we take the average number of working nodes in each period over the independent 100 runs, respectively. Moreover, we do the same with the dynamic working probability shown in Fig. 6.3. Simulation results for constant working probabilities and dynamic working probabilities are plotted in Fig. 6.4.

As shown in Fig. 6.4, with the dynamic probability, the number of working nodes attains the required number (namely, 200) within the first 29 periods. This is in line with the previous analysis. From the 30th period onward, the number of working nodes decreases quickly since the number of effective nodes is fewer than 200 as shown in Fig. 6.2. If the working probability is fixed to 0.2 or 0.3, from the 15th or 18th period onward, the number of working nodes falls below the required number of 200.

The simulation results demonstrate that the dynamic working probability delivered by Algorithm 6.1 makes the network performance more stable and prolongs the network lifetime as well.

## 6.3 Stochastic scheduling considering node failure

In real applications, sensor nodes may suffer internal faults or external damage caused by harsh environments. After a WSN is deployed, nodes are unlikely to be maintained. When a network starts running, any node may become defunct at any time.

### 6.3.1 Model of node failure

For a sensor node, both internal faults and external damages are difficult to predict. Hence we formulate them with a random process.

Suppose that, within a period length $T$, the probability for a failure happening on an effective node is $\lambda$. Denote the number of failures occurring within this period as $F_i$. As the number of effective nodes at the beginning of the $i$th period is $N_i$, $F_i$ follows the binomial distribution $B(N_i, \lambda)$, i.e. the probability that $F_i = n_i$ is

$$\text{Prob}(F_i = n_i) = \left( \begin{array}{c} N_i \\ n_i \end{array} \right) \lambda^{n_i} (1 - \lambda)^{N_i - n_i}. \tag{6.14}$$

Hence the expectation of $F_i$ is

$$\text{E}(F_i) = \lambda N_i. \tag{6.15}$$

### 6.3.2 Dynamic turning of working probability

Based on the analysis in Section 6.2.2, to maintain a stable performance, the working probability in the current period for each node is dependent on the number of effective nodes at the beginning of this period in the network.

As each node operates independently, within every period, it may become ineffective due to failure or energy exhaustion. Hence we first exclude the ineffective nodes which have used up energy, and then exclude the failure nodes. Then the remaining nodes are effective nodes and their working probability can be determined.

If node failure does not occur, the working probability for the 1st period is $\rho_1 = M_0/N$, where $M_0$ is the number of working nodes for the network performance and $N$ is the total number of nodes in the network. All nodes are assumed effective initially.

Denote by $\tilde{N}_i$ the number of effective nodes at the beginning of the $i$th period; hence $\tilde{N}_1 = N$. Denote by $F_i$ the number of node failures happening within the $i$th period. Denote by $\tilde{\rho}_i$ the working probability for each node in the $i$th period.

With Eq. (6.15), it is expected that there would be $E(F_1) = \lambda \tilde{N}_1$ nodes showing failure during the 1st period. Hence, to compensate for the loss caused by defunct nodes on the network performance, the working probability for each node in the 1st period should be $\tilde{\rho}_1$, namely

$$\tilde{\rho}_1 = \frac{M_0}{\tilde{N}_1 - E(F_1)} = \frac{M_0}{(1 - \lambda)\tilde{N}_1}. \tag{6.16}$$

Moreover, the expected number of effective nodes at the beginning of the 2nd period (or at the end of the 1st period), $\tilde{N}_2$, is

$$E(\tilde{N}_2) = \phi_1[\tilde{N}_1 - E(F_1)] = \phi_1(1 - \lambda)\tilde{N}_1, \tag{6.17}$$

where $\phi_1$ is the probability that each node is effective at the end of the 1st period. We have discussed how to compute this probability in detail in Section 6.2.2 (see e.g. Eq. (6.11)).

So, the expected number of failures within the 2nd period, $E(F_2)$, is

$$E(F_2) = \lambda E(\tilde{N}_2). \tag{6.18}$$

Taking into account the possible failures, the working probability in the 2nd period, $\tilde{\rho}_2$, is

$$\tilde{\rho}_2 = \frac{M_0}{E(\tilde{N}_2) - E(F_2)} = \frac{M_0}{(1 - \lambda)E(\tilde{N}_2)}. \tag{6.19}$$

The expected number of effective nodes at the beginning of the 3rd period, $E(\tilde{N}_3)$, is

$$E(\tilde{N}_3) = \phi_2[E(\tilde{N}_2) - E(F_2)] = \phi_2(1 - \lambda)\tilde{N}_2. \tag{6.20}$$

In general, the working probability in the $m$th period, $\tilde{\rho}_m$, is

$$\tilde{\rho}_m = \frac{M_0}{(1 - \lambda)E(\tilde{N}_m)}, \quad m = 1, 2, \cdots. \tag{6.21}$$

And the expected number of effective nodes at the beginning of the $(m + 1)$th period, $E(\tilde{N}_{m+1})$, is

$$E(\tilde{N}_{m+1}) = \phi_m(1 - \lambda)E(\tilde{N}_m), \quad m = 1, 2, \cdots. \tag{6.22}$$

The scheduling algorithm for the dynamic working probability considering node failure is as follows.

**Algorithm 6.2** (Computing the dynamic working probability considering node failure).

Step 1: Computing the number of working nodes, $M_0$, according to the requirement on the network performance with Eq. (6.2); determining the initial working probability with Eq. (6.16); setting $m = 1$.

Step 2: with Eq. (6.17) (or Eq. (6.22)), computing the expected number of effective nodes after $m$ periods; computing the working probability for the $(m + 1)$th period with Eq. (6.21); $m = m + 1$.

Step 3: repeating Step 2.

Different from Algorithm 6.1, this algorithm eliminates the impact by failure nodes, making the network performance stable.

## 6.4 Performance improvement

In Sections 6.2 and 6.3, the working probability for the current period hinges on the number of effective nodes at the end of last period. However, within the current period, nodes may exhaust energy to become ineffective and affect the network performance. In this section, we discuss the disadvantage of Algorithm 6.1 in Section 6.2 and propose a revision to the algorithm. Moreover, we investigate the relationship between the period length and the network lifetime and then develop an algorithm to optimize the period length.

### 6.4.1 Revised dynamic working probability

Based on the analysis in Section 6.2, we denote the working probabilities for the first $m(m > 1)$ periods by $\rho_1, \rho_2, \cdots, \rho_m$, respectively. Then the probability that a node is effective at the beginning of the $(m + 1)$th period can be obtained by Eq. (6.11).

Suppose a working node runs out of energy within the $(m + 1)$th period; then its energy at the beginning of this period is less than $E_{ww}$. Hence we first calculate the number of nodes which have more energy than $E_{ww}$ at the end of the $m$th period and then compute the working probability for the $(m + 1)$th period to see if we have enough working nodes.

We redefine an effective node as a node which has more energy than $E_{ww}$. Recall that in Section 6.1 an effective node is defined as a node which has not used up its energy. The difference of these two definitions is the amount of energy $E_{ww}$. We revise Eq. (6.9) by replacing the initial energy $E_o$ with $E_o - E_{ww}$, that is, setting the maximal number of working periods in the first $m$ periods as $\max\{u_m\} - 1$. Hence the probability that every node is effective at the end of the $m$th period, denoted $\hat{\phi}_m$, is

$$\hat{\phi}_m = \sum_{\substack{m \\ \sum_{i=1} v_i \le \max\{u_m\} - 1}} \prod_{j=1}^{m} q(v_j), \qquad (6.23)$$

where $\max(u_m)$ and $q(v_i)$ are determined by Eq. (6.9) and Eq. (6.10), respectively.

Similar to Eq. (6.12) and Eq. (6.13), the expected number of effective nodes at the beginning of the $(m+1)$th period, denoted $E(\hat{N}_{m+1})$, is

$$E(\hat{N}_{m+1}) = N_m \hat{\phi}_m, \tag{6.24}$$

and the revised working probability for the $(m+1)$th period, denoted as $\hat{\rho}_{m+1}$, is

$$\hat{\rho}_{m+1} = \frac{M_0}{E(\hat{N}_{m+1})}. \tag{6.25}$$

The procedure for computing the revised working probability for each period is similar to Algorithm 6.1.

**Algorithm 6.3** (Revising dynamic working probability).

Step 1: Computing the number of working nodes according to the requirement on the network performance (e.g. Eq. (6.2)); determining the initial working probability with Eq. (6.4); setting $m = 1$.

Step 2: With Eq. (6.23) and Eq. (6.24), computing the expected number of effective nodes at the end of $m$th period; with Eq. (6.25), calculating the working probability for the $(m+1)$th period; $m = m + 1$.

Step 3: Repeating Step 2.

Using the same parameters as listed in Table 6.1, we plot the working probability obtained with Algorithm 6.1 and the revised working probability with Algorithm 6.3 in Fig. 6.5.

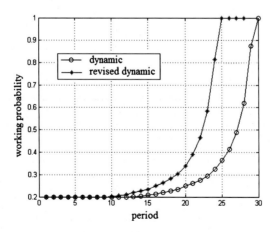

**FIGURE 6.5**

Working probability and revised working probability.

Fig. 6.5 shows that, as the number of periods increases and part of the nodes run out of energy, the working probability is smaller than the revised one, and their gap enlarges gradually.

### 6.4.2 **Optimization of the period length**

Every period starts with a configuration phase, in which each node will decide on its state with the probability in the remaining time of the period (see Fig. 6.1). Usually, the time $t_s$ is basically determined by the node's physical properties. Compared to $t_s$, the period length, $T$, should be big enough to reduce the amount of energy spent in the configuration phase. On the other hand, a shorter $T$ means less energy cost within a period and hence is helpful to balance energy consumption among sensor nodes and prolong the network lifetime. Suppose $t_s$ and $E_o$ are given. Then the network lifetime is largely dependent on the value of $T$.

The maximum period length, $T_{max}$, cannot exceed the maximal working time, $E_o/Q_{on}$. Hence we have $T \in (t_s, E_o/Q_{on}]$. As $t_s$ is short, we use $t_s$ as the step size in iterations to find the best value of $T$, which can meet the requirement on the network performance and prolong the expected network lifetime as well.

The algorithm for optimizing $T$ is as follows.

**Algorithm 6.4** (Optimizing the period length).

Step 1: Computing the number of working nodes, $M_0$, according to the requirement on the network performance and initialize $L_{max} = 0$, $T_{best} = 0$, $T_{max} = E_o/Q_{on}$, $T = 2t_s$, $m = 1$.

Step 2: Computing working probability with Algorithm 6.1 and, with Eq. (6.7) or Eq. (6.12), computing $E(N_{m+1})$.

Step 3: If $E(N_{m+1}) \geq M_0$, then $m = m + 1$ goto Step 2; otherwise, goto Step 4.

Step 4: If $L_{max} < mT$, then $L_{max} = mT$, $T_{best} = T$.

Step 5: If $T \geq T_{max}$, goto Step 6; otherwise, $T = T + t_s$, $m = 1$ and goto Step 2.

Step 6: Termination.

Algorithm 6.4 lets $T$ go through the interval $(t_s, E_o/Q_{on})$ with $t_s$ as its step size and calculates the expected network lifetime for each $T$ to find the optimal period length, $T_{best}$.

## 6.5 **Summary**

In this chapter, to maintain the network performance stability and prolong the network lifetime, a scheduling mechanism is proposed to make the number of working nodes keep at the same level within each period. An algorithm is developed to determine the working probability based on the number of effective nodes. Simulation results validate the effectiveness of the algorithm. Considering the possibility of node failure, a revision of the dynamic working probability is necessary to compensate for the loss due to defunct nodes. The relationship between the period length and the network lifetime is analyzed and an algorithm is presented to optimize the length.

# References

[1] H. Bai, X. Chen, Y.C. Ho, X. Guan, Percentage coverage configuration in wireless sensor networks, in: ISPA 2005, in: LNCS, vol. 3758, 2005, pp. 780–791.

[2] H. Bai, X. Chen, B. Li, D. Han, A location-free algorithm of energy-efficient connected coverage for high density wireless sensor networks, Discrete Event Dynamic Systems 17 (1) (March 2007) 1–21.

[3] J. Li, X. Chen, Research of stochastic scheduling algorithm for wireless sensor networks (in Chinese), Journal of Computer Applications 31 (3) (2011) 594–597, 605.

[4] C. Schurgers, V. Tsiatsis, S. Ganeriwal, et al., Optimizing sensor networks in the energy-latency density design space, IEEE Transactions on Mobile Computing 1 (1) (2002) 70–80.

# Energy-based multisource localization

## CONTENTS

Wireless sensors can collect the information and transmit data of targets for localization. Detection and tracking of targets are important applications of WSNs.

Acoustic signals are widely used for source localization. Various methods for acoustic source localization with microphone arrays in different environments are proposed [2,4,7,12,20,22]. Physical measurements for acoustic source are classified into three types: time delay of arrival (TDOA) [2,14,19,24], received signal strength indication (RSSI) [10,11,17], and direction of arrival (DOA) [5,8,13,21]. Since the acoustic energy emitted by sources usually varies slowly, compared to the raw acoustic time series, the acoustic energy time series can be sampled at a lower rate [17] and few data need to be transmitted to the fusion center. Therefore, sensor nodes can save energy on data transmissions and communication bandwidth over shared wireless channels.

In real applications, there are usually multiple sources in the area of interest but the source number is unknown. Compared to single source localization, the key challenge for multisource localization is how to determine the source number and locations based on the superimposed signals.

**Randomly Deployed Wireless Sensor Networks. https://doi.org/10.1016/B978-0-12-819624-3.00012-4**
Copyright © 2020 Tsinghua University Press. Published by Elsevier Inc. All rights reserved.

In this chapter, the problem of source localization based on acoustic signals is considered. In Section 7.1, multisource localization with a given source number is studied. The formulation of a maximum likelihood estimation is presented and the methodology for solving this problem is discussed. Then in Section 7.2, the source number estimation is investigated. Two algorithms based on clustering and the minimum description length are developed.

The assumptions imposed in this chapter are as follows.

**Assumption 7.1.** *The area of interest is a plane region without any obstacle.*

**Assumption 7.2.** *Multipath effects in acoustic propagation are ignored.*

**Assumption 7.3.** *Each sensor node knows its own location.*

**Assumption 7.4.** *Each sensor node has an unique ID.*

Under these assumptions, the results obtained in this chapter will provide an opportunity to extend the proposed solution to real applications.

## 7.1 Multisource localization with given source number

Multisource localization is essentially by estimating multiple parameters based on observation information. These parameters are the source number, source locations and other information about sources. In this section, we first consider a simple situation in which the source number is given.

### Signal energy observation model

Suppose there are $N$ sensor nodes (microphones) in the network, and there are $K$ acoustic sources in the area of interest; both $N$ and $K$ are given. Denote the location of node $O_i$ as $\gamma_i$ ($i = 1, 2, \cdots, N$), which is given; and the location of the $k$th source, $\zeta_k(t)$ ($k = 1, \cdots, K$), is to be estimated [6,9,17,18]. Denote

$$d_{ik}(t) = \| \zeta_k(t) - \gamma_i \| \tag{7.1}$$

as the Euclidean distance from $O_i$ to the $k$th source.

Suppose that the acoustic energy measured at node $O_i$ over a fixed time duration $t$ is (see Equation (4) in [17])

$$z_i(t) = g_i \sum_{k=1}^{K} \frac{U_k(t)}{d_{ik}^2(t)} + \epsilon_i(t), \ i = 1, 2, \cdots, N, \tag{7.2}$$

where $g_i$ is the gain factor of $O_i$. $g_i$ is assumed to be known by using sensor calibration. $U_k(t)$ is the energy strength at 1 meter away from the $k$th source. $U_k(t)$ needs to be estimated. $\epsilon_i(t)$ represents the measurement variation, which is approximated with a normal distribution, namely $\epsilon_i(t) \sim N(\mu_i, \sigma_i^2)$.

### 7.1.1 Localization using maximum likelihood estimation

Sheng and Hu [17] present a maximum likelihood acoustic source location estimation method for wireless ad hoc sensor networks with the number of sources being given. We summarize the method in the following.

For simplicity, $t$ may be omitted in Eq. (7.2) and its matrix form is as follows:

$$y = (y_1 \ y_2 \ \cdots \ y_N)^{\mathrm{T}} = \left( \frac{z_1 - \mu_1}{\sigma_1} \ \frac{z_2 - \mu_2}{\sigma_2} \ \cdots \ \frac{z_N - \mu_N}{\sigma_N} \right)^{\mathrm{T}},$$

$$G = \begin{bmatrix} \frac{g_1}{\sigma_1} & 0 & \mathbf{0} & 0 \\ 0 & \frac{g_2}{\sigma_2} & \mathbf{0} & 0 \\ \mathbf{0} & \mathbf{0} & \ddots & \mathbf{0} \\ 0 & 0 & \mathbf{0} & \frac{g_N}{\sigma_N} \end{bmatrix},$$

$$D = \begin{bmatrix} \frac{1}{d_{11}^2} & \frac{1}{d_{12}^2} & \cdots & \frac{1}{d_{1K}^2} \\ \frac{1}{d_{21}^2} & \frac{1}{d_{22}^2} & \cdots & \frac{1}{d_{2K}^2} \\ \vdots & \vdots & \ddots & \vdots \\ \frac{1}{d_{N1}^2} & \frac{1}{d_{N2}^2} & \cdots & \frac{1}{d_{NK}^2} \end{bmatrix},$$

$$U = (U_1 \ U_2 \ \cdots \ U_K)^{\mathrm{T}},$$

$$\upsilon = (\upsilon_1 \ \upsilon_2 \ \cdots \ \upsilon_N)^{\mathrm{T}} = \left( \frac{\epsilon_1 - \mu_1}{\sigma_1} \ \frac{\epsilon_2 - \mu_2}{\sigma_2} \ \cdots \ \frac{\epsilon_N - \mu_N}{\sigma_N} \right)^{\mathrm{T}},$$

(7.3)

where the operation $a^{\mathrm{T}}$ means the transpose of vector $a$; $\upsilon_i \sim N(0,1)$ $(i = 1, 2, \cdots, N)$ is uncorrelated noise. Eq. (7.2) may be written as

$$y = GDU + \upsilon. \tag{7.4}$$

The joint distribution of $y$ is

$$f(y|\psi) = \frac{1}{(\sqrt{2\pi})^N} \exp\{-\frac{1}{2}(y - GDU)^{\mathrm{T}}(y - GDU)\}, \tag{7.5}$$

where $\psi = [\zeta_1^{\mathrm{T}}, \zeta_2^{\mathrm{T}}, \cdots, \zeta_K^{\mathrm{T}}, U_1, U_2, \cdots, U_K]^{\mathrm{T}}$. Note that $\zeta_k$ and $U_k$ are the location and signal energy strength of the $k$th source. Define

$$e(\psi) = (y - GDU)^{\mathrm{T}}(y - GDU) = \| y - GDU \|^2. \tag{7.6}$$

$e(\psi)$ measures the estimate error with respect to $\psi$. The localization problem is equivalent to minimizing $e(\psi)$ and also equivalent to maximizing $f(y|\psi)$ defined in Eq. (7.5).

With some more derivation steps (see [17] for details), to minimize $e(\psi)$, the estimate of location $\zeta_k$ is a linearly weighted sum of node locations, namely

$$\zeta_k = \frac{\sum_{i=1}^{N} w_{ik} \gamma_i}{\sum_{i=1}^{N} w_{ik}}, \quad k = 1, 2, \cdots, K, \tag{7.7}$$

where

$$w_{ik} = \frac{g_i}{\sigma_i} \left( \frac{1}{d_{ik}^2} \right) \left( z_i - \frac{g_i}{\sigma_i} \sum_{k=1}^{K} \frac{U_k}{d_{ik}^2} \right).$$

Eq. (7.7) is not an explicit expression for $\zeta_k$ because $w_{ik}$ depends on the distance $d_{ik}$ and $d_{ik}$ depends on $\zeta_k$.

## 7.1.2 Location initialization for localization

If the initial point is close to the source location, search algorithms will avoid being trapped in a local minimum and reduce the computation complexity. In this section, a method for source location initialization based on Voronoi diagrams is presented.

Suppose there are $N$ sensor nodes in a plane. If the plane is partitioned into $N$ convex polygons such that each polygon contains exactly one sensor node and every point in a given polygon is closer to the node in this polygon than to any other node, then we get a Voronoi diagram. In a Voronoi diagram, each polygon is called Voronoi cell (also see Section 3.1.1).

In the Voronoi diagram, if node $O_j$ and node $O_i$ share a Voronoi edge, then $O_j$ is a neighbor of $O_i$. Denote $\Pi(i)$ as the set of all neighbors of $O_i$.

### *Single source localization*

Consider the case when there is only one source in the area of interest. Suppose its location and energy strength are $\zeta_0 = (\zeta_0^x, \zeta_0^y)$ and $u$, respectively. According to the definition of a Voronoi cell, for any node located within the area of interest, it has at least three neighbors. For some sensor node, say $O_{i_0}$, we arbitrarily choose three neighbors $O_{i_1}$, $O_{i_2}$ and $O_{i_3}$ from $\Pi(i_0)$. Suppose the energy readings of these four nodes are $z_{i_0}$ and $z_{i_1}, z_{i_2}, z_{i_3}$, and their locations are $\gamma_{i_0}$ and $\gamma_{i_1}, \gamma_{i_2}, \gamma_{i_3}$, respectively. In Eq. (7.2), if ignoring the noise, then

$$z_{i_h} = g_{i_h} \frac{u}{\| \zeta_0 - \gamma_{i_h} \|^2}, \quad h = 0, 1, 2, 3. \tag{7.8}$$

Hence

$$\| \zeta_0 \|^2 + \| \gamma_{i_h} \|^2 - 2\zeta_0^{\mathrm{T}} \gamma_{i_h} = g_{i_k} \frac{u}{z_{i_h}}, \quad h = 0, 1, 2, 3. \tag{7.9}$$

Subtracting Eq. (7.9) when $h = 0$ from Eq. (7.9) when $h = 1, 2, 3$, we obtain

$$
\begin{aligned}
2(\gamma_{i_1} - \gamma_{i_0})^{\mathrm{T}}\zeta_0 + u(g_{i_1}/z_{i_1} - g_{i_0}/z_{i_0}) &= \| \gamma_{i_1} \|^2 - \| \gamma_{i_0} \|^2, \\
2(\gamma_{i_2} - \gamma_{i_0})^{\mathrm{T}}\zeta_0 + u(g_{i_2}/z_{i_2} - g_{i_0}/z_{i_0}) &= \| \gamma_{i_2} \|^2 - \| \gamma_{i_0} \|^2, \\
2(\gamma_{i_3} - \gamma_{i_0})^{\mathrm{T}}\zeta_0 + u(g_{i_3}/z_{i_3} - g_{i_0}/z_{i_0}) &= \| \gamma_{i_3} \|^2 - \| \gamma_{i_0} \|^2.
\end{aligned}
\tag{7.10}
$$

Let

$$
C = \begin{pmatrix}
2(\gamma_{i_1} - \gamma_{i_0})^{\mathrm{T}} & g_{i_1}/z_{i_1} - g_{i_0}/z_{i_0} \\
2(\gamma_{i_2} - \gamma_{i_0})^{\mathrm{T}} & g_{i_2}/z_{i_2} - g_{i_0}/z_{i_0} \\
2(\gamma_{i_3} - \gamma_{i_0})^{\mathrm{T}} & g_{i_3}/z_{i_3} - g_{i_0}/z_{i_0}
\end{pmatrix}, \quad
d = \begin{pmatrix}
\| \gamma_{i_1} \|^2 - \| \gamma_{i_0} \|^2 \\
\| \gamma_{i_2} \|^2 - \| \gamma_{i_0} \|^2 \\
\| \gamma_{i_3} \|^2 - \| \gamma_{i_0} \|^2
\end{pmatrix}.
$$

The matrix form of Eq. (7.10) is

$$
C \begin{pmatrix} \zeta_0^x \\ \zeta_0^y \\ u \end{pmatrix} = d.
\tag{7.11}
$$

As the three neighbors of $O_{i_0}$ are not collinear, $C$ is invertible. Hence

$$
\begin{pmatrix} \zeta_0^x \\ \zeta_0^y \\ u \end{pmatrix} = C^{-1}d.
\tag{7.12}
$$

### *Multisource location initialization*

Intuitively, if one node's energy reading is larger, then it is closer to one source. Therefore we select the node which has a larger reading among its neighbors. By assuming that the readings of the node and its neighbors are all from a single source, we estimate its location and energy strength. The algorithm for multisource location initialization is as follows.

**Algorithm 7.1** (Multisource location initialization with Voronoi diagram).
   Input: Source number $K$, energy readings $z_1, z_2, \cdots, z_N$; node locations $\gamma_1, \gamma_2, \cdots, \gamma_N$.
   Output: $\hat{\zeta}_{01}, \cdots, \hat{\zeta}_{0K}$ as initial source locations.

   Step 1: Generating a Voronoi diagram with $\gamma_1, \gamma_2, \cdots, \gamma_N$. Selecting the node whose reading is the biggest one as the first node $O_{n_1}$, denote $\Lambda = \{n_1\}$. Constructing $\mathbf{C}_1$ and $d_1$ in Eq. (7.11) using $O_{n_1}$ and its three noncollinear neighbors. With Eq. (7.12), estimating location and energy strength of the first source, $(\hat{\zeta}_{01}^{\mathrm{T}}, \hat{U}_1)^{\mathrm{T}} = \mathbf{C}_1^{-1}d_1$. Setting $K' = 2$.

Step 2: Removing the effect of the estimated source(s) from the readings of all nodes

$$z'_i = z_i - g_i \sum_{k=1}^{K'-1} \frac{\hat{U}_k}{\| \hat{\zeta}_{0k} - \gamma_i \|^2}, \quad i = 1, 2, \cdots, N,$$

$$n_{K'} = \arg \max_{i \in \{1,2,\cdots,N\} \backslash \Lambda} \{z'_i\}.$$

Step 3: Computing $\sigma_{n_{K'}}$, which is the standard deviation of the readings of node $O_{n_{K'}}$ and its neighbors.

Step 4: If $z'_{n_{K'}} - \dfrac{\sum_{j \in \Pi(n_{K'})} z'_j}{|\Pi(n_{K'})|} > 3\sqrt{1 + \dfrac{1}{|\Pi(n_{K'})|}} \sigma_{n_{K'}}$ where $|\Pi(n_{K'})|$ represents the number of elements in set $\Pi(n_{K'})$, i.e. $|\Pi(n_{K'})|$ is the number of neighbors, then

$$\Lambda = \Lambda \cup \{n_{K'}\},$$

using Eq. (7.12) to estimate the location $\hat{\zeta}_{0K'}$ and energy strength $\hat{U}_{K'}$ of the $K'$th source;

$K' = K' + 1$. If $K' < K$, going to Step 2; otherwise, going to Step 7.

Step 5: If $z'_{n_{K'}} - \dfrac{\sum_{j \in \Pi(n_{K'})} z'_j}{|\Pi(n_{K'})|} \leq 3\sqrt{1 + \dfrac{1}{|\Pi(n_{K'})|}} \sigma_{n_{K'}}$, selecting the node whose reading is the second largest among its neighbors, denoting it as node $n_{K'}$, going to Step 3; if no such node exists any more, going to Step 6.

Step 6: Selecting $K - K' + 1$ locations from $\{\hat{\zeta}_{01}, \hat{\zeta}_{02}, \cdots, \hat{\zeta}_{0(K'-1)}\}$ and denoting them as $\hat{\zeta}_{0K'}, \cdots, \hat{\zeta}_{0K}$.

Step 7: Returning initial source location $\hat{\zeta}_{01}, \cdots, \hat{\zeta}_{0K'}$. Termination.

When there are two or more sources located in one Voronoi cell, they are viewed as one source with this algorithm. Hence we introduce triple standard deviation condition to identify the node which is located closest to some source. When a node's reading is the biggest one, and the reading difference with each of its neighbors is more than the triple standard deviation, this node is located closest to some source; if the triple standard deviation condition is not satisfied, most likely the remaining $K - K' + 1$ sources are close to sources which have been initialized and thus we take Step 6 to complement the initialization.

### *Multiresolution search based on source location initialization*

Starting from the initial locations obtained with Algorithm 7.1, we use a multiresolution search to find more accurate source locations. In fact, source location initialization may be viewed as the earlier iterations of the multiresolution search algorithm. We implement a multiresolution search around the initial locations until the resolution is the same as with the regular (without location initialization) multiresolution search.

Suppose the region on each dimension of the area of interest is $L$ meters in length, which is divided into $q$ grids. The region for a multiresolution search is $\delta$ meters around the initial location. There are $w$ points to be searched per iteration, hence the number of iterations, $m'$, satisfies $w^{m'} = q\delta/L$, namely $m' = \log_w(q\delta/L)$. This is to be compared to the regular multiresolution search in which the iteration number $m = \log_w q$, $m'$ is smaller than $m$. For example, when $L = 100$ and $q = 1024$, the minimum resolution, $L/q$, is about 0.1. Suppose $\delta = 10$ and $w = 2$, then $m = 10$ and $m' \approx 7$. The number of iterations decreases from 10 to 7. The multiresolution search algorithm with source location initialization is summarized as follows.

**Algorithm 7.2** (Multiresolution search with source location initialization).

Input: $\delta$, $L$, $w$, $q$; energy readings $z_1, z_2, \cdots, z_N$; node locations $\gamma_1, \gamma_2, \cdots, \gamma_N$.
Output: the estimate of source locations $\hat{\zeta}_1, \hat{\zeta}_2, \cdots, \hat{\zeta}_K$.

Step 1: With Algorithm 7.1, initializing source locations
$\hat{\zeta}_{01} = (\hat{\zeta}_{01}^x, \hat{\zeta}_{01}^y)$, $\hat{\zeta}_{02} = (\hat{\zeta}_{02}^x, \hat{\zeta}_{02}^y)$, $\cdots$, $\hat{\zeta}_{0K} = (\hat{\zeta}_{0K}^x, \hat{\zeta}_{0K}^y)$.

Step 2: Calculating $m' = \log_w \dfrac{\delta q}{L}$;
The search scope for the $k$th source is

$$G_{0,k} = \left\{ (g_x, g_y): \quad g_x \in \{\hat{\zeta}_{0k}^x - \delta/2 + \delta/w, \hat{\zeta}_{0k}^x - \delta/2 + 2\delta/w, \cdots, \hat{\zeta}_{0k}^x + \delta/2\}, \right.$$
$$\left. g_y \in \{\hat{\zeta}_{0k}^y - \delta/2 + \delta/w, \hat{\zeta}_{0k}^y - \delta/2 + 2\delta/w, \cdots, \hat{\zeta}_{0k}^y + \delta/2\} \right\},$$

where $k = 1, 2, \cdots, K$.
Let $i = 1$.

Step 3: From $G_{i-1,k}(k = 1, 2, \cdots, K)$, choosing the locations of $K$ sources which minimize Eq. (7.6) and denoting them as $\hat{\zeta}_{i1}, \cdots, \hat{\zeta}_{iK}$, respectively.

Step 4: Adjusting the search scope of the $k$th source as

$$G_{i,k} = \left\{ (g_x, g_y): \quad g_x \in \{\hat{\zeta}_{ik}^x - \frac{\delta}{2} + \frac{\delta}{w}, \hat{\zeta}_{ik}^x - \frac{\delta}{2} + \frac{2\delta}{w}, \cdots, \hat{\zeta}_{ik}^x + \frac{\delta}{2}\}, \right.$$
$$\left. g_y \in \{\hat{\zeta}_{ik}^y - \frac{\delta}{2} + \frac{\delta}{w}, \hat{\zeta}_{ik}^y - \frac{\delta}{2} + \frac{2\delta}{w}, \cdots, \hat{\zeta}_{ik}^y + \frac{\delta}{2}\} \right\},$$

where $k = 1, 2, \cdots, K$.
If $i + 1 \geq m'$, going to Step 5; otherwise, setting $\delta = \delta/w$ and $i = i + 1$, going to Step 3.

Step 5: Setting $\hat{\zeta}_k = \hat{\zeta}_{ik}$, $k = 1, 2, \cdots, K$. Returning the estimate of source locations $\hat{\zeta}_1, \hat{\zeta}_2, \cdots, \hat{\zeta}_K$.

### 7.1.3 Simulation and analysis

We simulate and analyze Algorithm 7.2. We test its performance under different number of nodes and compare it with a regular multiresolution search algorithm. The localization error is defined as $\sum_{k=1}^{K} \| \hat{\zeta}_k - \zeta_k \| / K$, where $K$ is the source number,

$\hat{\zeta}_k$ and $\zeta_k$ are the estimated location and the real location of the $k$th source, respectively.

We randomly and uniformly distribute $N$ nodes in the area of $100 \times 100$ square meters. For node $O_i$, its gain factor $g_i = 1$ $(1 \leq i \leq N)$, and its measurement variation $\epsilon_i$ is a Gaussian random variable in $N(1, 1/50)$. Two sources are located at $(25, 50)$ and $(75, 50)$, as shown in Fig. 7.1. The energy strength for each source is independently randomly generated following the uniform distribution $U[3000, 7000]$.

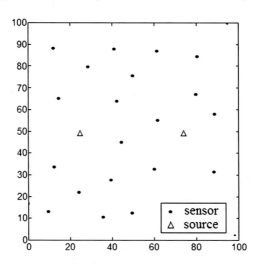

**FIGURE 7.1**

Locations of sensor nodes and sources.

The following experiments aim to examine the impact of the number of nodes on the localization error. We set $q = 100$, $w = 3$, and $\delta = 10$ meters for multiresolution search. We take the number of nodes $N = 20, 30, 40, 50, 60, 70, 80, 90$, and $100$. For each value of $N$, we have 100 independent replications. In each one, every node's location is randomly generated. Finally, over the 100 replications, we get the average localization errors. The initial locations for regular multiresolution search are arbitrarily fixed. The regular multiresolution search is likely trapped in a local minimum. With initial source locations based on Voronoi diagrams, a multiresolution search yields more accurate estimates, as shown in Fig. 7.2.

When the number of nodes $N \geq 30$, the source localization error is below 5 meters. More nodes help to reduce the error. When there are fewer sensor nodes in the network, on average, nodes are relative far away from sources so that signal-to-noise ratio is lower and localization error becomes bigger.

Based on the initial source locations obtained by Algorithm 7.1, Algorithm 7.2 helps to improve localization accuracy by searching around the initial source locations with higher resolution. Fig. 7.3 illustrates the updating process. The estimated location gradually moves closer to the real location, $(25, 50)$.

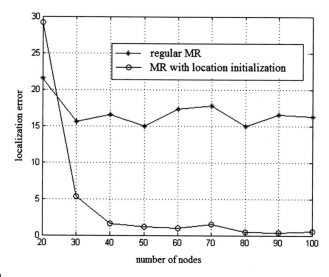

**FIGURE 7.2**

Localization error (MR: multiresolution).

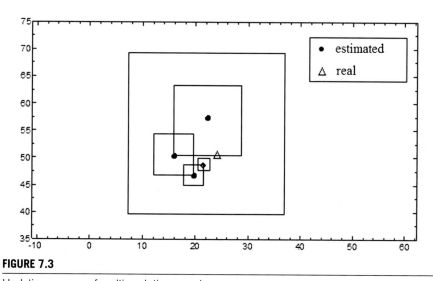

**FIGURE 7.3**

Updating process of multiresolution search.

Fig. 7.4 shows the updated results with the multiresolution search. When the number of nodes $N \geq 30$, the error of the initial source locations obtained by Algorithm 7.1 is below 5 meters; a multiresolution search based on the initial locations reduces the error by about 10%.

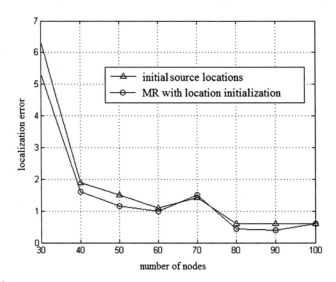

**FIGURE 7.4**

Errors of initial source locations and results of multiresolution (MR) search.

## 7.2 Source number estimation

In real applications, the source number may be unknown. But it is crucial for localization. In this section, we consider the problem of an energy-based multisource number estimation. The energy measurement model (see Eq. (7.2)) in Section 7.1 is

$$z_i(t) = g_i \sum_{k=1}^{K} \frac{U_k(t)}{d_{ik}^2(t)} + \epsilon_i(t), \; i = 1, 2, \cdots, N. \tag{7.13}$$

Now the source number $K$ is unknown and also needs to be estimated in the problem of multisource localization. As $K$ defines the form of $z_i(t)$ in Eq. (7.13), the maximum likelihood estimation method is not applicable if $K$ is unknown.

We present two methods to estimate the source number. One is an intuitive method based on senor clustering [3]. As sensor nodes which are close to the sources will have bigger energy strength measurements, we first select such sensor nodes, then make the cluster accordingly and use the number of clusters as the estimate of the source number. The other one applies the Minimum Description Length (MDL) method to estimate the source number.

To effectively distinguish different sources, we further assume that the distance between any two sources is big enough. If two sources are too close, due to the superimposed energy strength measurement model, most likely they are treated as a single source.

When estimating the source number, nodes will forward their energy readings to the sink node or the base station and data processing will be carried out in a centralized way.

In the same way as in Section 7.1, suppose the WSN consists of $N$ nodes and there are $K$ sources in the area of interest ($K$ is unknown). $O_i$'s location, $\gamma_i = (\gamma_i^x, \gamma_i^y)^T (i = 1, 2, \cdots, N)$, is given. Both the location and the energy strength of the $k$th source, $\zeta_k = (\zeta_k^x, \zeta_k^y)^T$ and $U_k$ ($k = 1, 2, \cdots, K$), are unknown.

## 7.2.1 Estimation based on node selection and clustering

We select out the nodes which have bigger readings. Theses nodes are more likely located closer to the sources and form a number of clusters around sources. Therefore we view the number of clusters as an estimate of the source number.

### Node selection

We first generate a Voronoi diagram based on the node locations, then pick those nodes whose energy readings are relative high.

An intuitive idea is setting one common threshold. A node will be selected if its reading is above the threshold. However, different sources may have different energy strengths. If the threshold is big, sources with lower energy strengths may be ignored; on the other hand, if the threshold is small, nodes which are far away from the sources may be selected; neither of them is good for clustering. Hence we use the concept of a neighbor in the Voronoi diagram to select the nodes. If a node's energy reading is bigger than the readings of its neighbors, then this node is more likely to be closer to some source, or there exists a source in the node's Voronoi cell.

If nodes are densely deployed in the area of interest, energy readings of two nearby nodes may show little difference. In addition, a node's reading may be very big due to the presence of the measurement noise even if it is not close to any source.

As shown in Fig. 7.5, sensor nodes are uniformly deployed in the area of $100 \times 100$. There are three sources located at $(25, 25)$, $(50, 75)$ and $(75, 25)$, respectively. Sensor nodes whose readings are greater than the threshold are selected and marked with '$*$'; ordinary sensor nodes are marked with '$\cdot$', and sources are marked with '$\triangle$'. When the number of nodes $N = 30$, all sources fall in the Voronoi cells of the selected nodes (Fig. 7.5A). When $N = 50$, because of the measurement noise, one node is selected although it is far away from all sources (Fig. 7.5B). If there are more nodes in the network ($N = 100$ or $N = 200$), more nodes are selected but many of them are far away from any source (Fig. 7.5C and Fig. 7.5D). In this case, the estimate of the source number will be wrong.

To get a more accurate estimate as the node density is higher, we propose one way for node selection in multiple rounds. The basic idea is randomly dividing all sensor nodes into several groups and letting each node belong to one and only one group. For each group, we select nodes which are close to the sources by local comparison. Then we gather all selected nodes for clustering to estimate the source number.

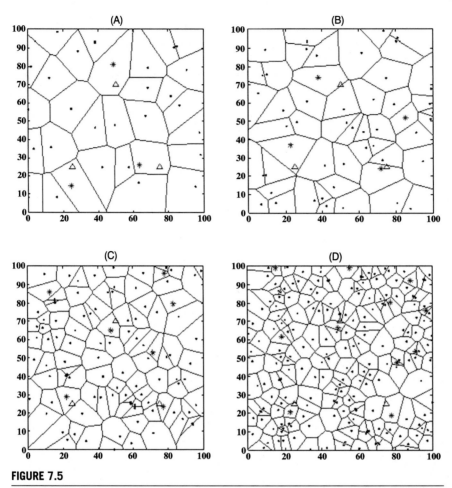

**FIGURE 7.5**

Impact of measurement noise on node selection. (A) $N = 30$. (B) $N = 50$. (C) $N = 100$.
(D) $N = 200$.

**Algorithm 7.3** (Multi-round local comparison).

Input: energy readings $z_1, z_2, \cdots, z_N$; node locations $\gamma_1, \gamma_2, \cdots, \gamma_N$.
Output: the set of selected nodes $\Gamma_c$.

Step 1: Randomly dividing all nodes into $J$ groups and denoting them as $G_1, G_2, \cdots, G_J$. Each group has $N/L$ nodes.
Setting $j = 1$.
Step 2: From $G_j$, selecting the nodes which have the biggest readings among their Voronoi neighbors and adding them in the set $\Gamma_c$.
Step 3: $j = j + 1$, if $j \le J$, going to Step 2; otherwise, returning the set of selected nodes $\Gamma_c$, termination.

To demonstrate the performance of Algorithm 7.3 by simulation, 400 sensor nodes are uniformly distributed in the area of $100 \times 100$ square meters. Three sources are located at $(25, 25)$, $(50, 75)$ and $(75, 25)$, respectively. The 400 nodes are divided into 20 groups and each of them has 20 nodes. Fig. 7.6 shows the selected nodes (marked with '$*$') yielded by multi-round local comparison. Obviously, the selected nodes cluster around the sources (marked with '$\triangle$') and the clusters separate distinctively. Hence we can estimate the source number correctly by the clustering nodes.

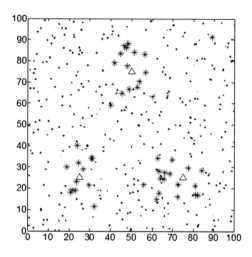

**FIGURE 7.6**

Selected nodes by multi-round local comparison.

### Node clustering

The set of selected nodes can be easily clustered by density-based spatial clustering [3]. First, we classify nodes into three types according to their locations. Denote by $\Upsilon$ the threshold for the number of neighbors and by $R_a$ the specified radius. A node is defined as a core node if it has more than $\Upsilon$ neighbors within the distance of $R_a$. A node is defined as a border node if it has fewer than $\Upsilon$ neighbors within $R_a$ but is in the neighborhood of a core node. If a node is neither core nor border node, then it is called a noise node.

Fig. 7.7 further explains the meaning of core node, border node and noise node. The algorithm for node clustering is as follows.

**Algorithm 7.4** (Node clustering).

Input: the set of selected nodes $\Gamma_c = \{\text{node}_1, \text{node}_2, \cdots, \text{node}_F\}$, $\Upsilon$, $R_a$.
Output: the number of clusters.

Step 1: Labeling each node in $\Gamma_c$ as core or border or noise senor according to their definitions.
Step 2: Eliminating noise nodes.

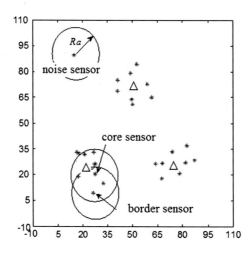

**FIGURE 7.7**

Illustration of core node, border node and noisy node ($\Upsilon = 3$, $R_a = 15$).

Step 3: Connecting core nodes if they are located within a distance of $R_a$.
Step 4. Letting each group of connected core nodes form a separate cluster.
Step 5. Assigning each border node to one cluster of its associated core nodes.
Step 6. Returning the number of clusters.

We view the number of clusters as an estimate of source number. $\Upsilon$ and $R_a$ are two key parameters for the estimate accuracy with Algorithm 7.4. We use a similar method to [3] to determine them. The idea is that, for nodes in a cluster, their $k$th nearest neighbors are at roughly the same distance, while noise nodes have the $k$th nearest neighbor at farther distance. Fig. 7.8 shows the selected nodes chosen from 400 nodes by 20 rounds. There is a leap in the distance of 5th nearest neighbor at 13 (marked with '$\times$') as shown in Fig. 7.9. So we set $\Upsilon$ to 5 and the radius $R_a$ to 13.

In fact, randomly dividing $N$ nodes into $J$ groups may result in different groups and deliver different estimates of the source number. To reduce the effect of randomness on the estimate, we run Algorithm 7.3 $I$ times and denote the corresponding estimates as $\hat{K}_1, \hat{K}_2, \cdots, \hat{K}_I$. Then we take the number which appears most frequently within these $I$ numbers as the estimate.

## 7.2.2 Estimation based on minimum description length criterion

Now we present an algorithm with minimum description length criterion to estimate the source number based on energy strength measurements. The source number estimation may be viewed as a model selection problem. As information theoretic criteria for model selection, both the Akaike information criterion introduced by Akaike [1] and the minimum description length criterion introduced by Rissanen [15,16] have proved very useful. The minimum description length criterion has been successfully

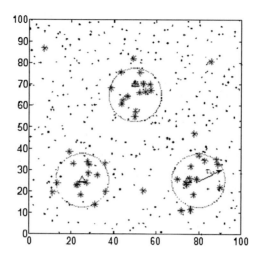

**FIGURE 7.8**

Selected nodes from 400 (20 × 20) nodes.

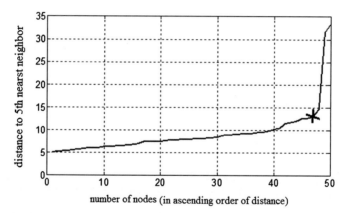

**FIGURE 7.9**

Distance to the 5th nearest neighbor for each node.

applied to detection of the number of signals received by a sensor array [23]. In this section, we discuss how to use the minimum description length criterion to estimate the source number based on energy measurements.

### *Minimum Description Length criterion (MDL)*

If the source number is $k$, according to the problem model in Section 7.1.1 and with Eqs. (7.3), (7.4), and (7.5), the maximum log-likelihood function is

$$F_k = F(\psi_k^*) = \max_{\psi_k} F(\psi_k) = \max_{\psi_k} \sum_{i=1}^{N} \ln f(y_i | \psi_k), \qquad (7.14)$$

where $\psi_k = [\zeta_1^T, \cdots, \zeta_k^T, U_1, \cdots, U_k]^T$.

Define the function MDL($k$) as follows:

$$\text{MDL}(k) = -F_k + \frac{3k}{2} \ln N, \qquad (7.15)$$

in which the second term is a penalty function which describes the model complexity.

Suppose $\bar{K}$ is an upper bound of the source number. Denote by $\hat{K}$ the best estimate of the source number, then

$$\hat{K} = \arg \min_{1 \le k \le \bar{K}} \{\text{MDL}(k)\}. \qquad (7.16)$$

Hence minimizing MDL($k$) means finding the best value of $k$ which makes log-likelihood function big enough and the model quite simple as well.

### Source parameters estimation with given source number

Given the source number $k$, to calculate the value of the maximum likelihood function $F_k$, we first estimate the locations and energy strength of the $k$ sources, namely $\hat{\psi}_k$. This is the problem of multisource localization with given source number. Of course, the maximum likelihood method in Section 7.1.1 is applicable but it is expensive in terms of computation time. As the estimation of $k$ is merely a middle step for solving Problem (7.16), one more convenient way is as follows.

Similar to the way in Section 7.2.1, the neighbor set for each node is built with a Voronoi diagram. Denote by $\Gamma_i$ the set of neighbors of $O_i$. Define the ratio of the measurement of $O_i$ to the average measurement of its neighbors as $w_i$ (measurement ratio for short), that is,

$$w_i = \frac{|\Gamma_i| z_i}{\sum_{j \in \Gamma_i} z_j}, \quad W = \{w_i\}_{i=1}^N, \qquad (7.17)$$

where $|\Gamma_i|$ is the number of neighbors of $O_i$; $W$ is the set of all ratios.

If there are $k$ sources, we select $k$ nodes with the following algorithm.

**Algorithm 7.5** (Node selection).

Input: The set of measurement ratios $W = \{w_i\}_{i=1}^N$; node locations $\{\gamma_i\}_{i=1}^N$.

Output: The selected nodes $\{\text{node}_i\}_{i=1}^k$.

Step 1: Arranging the elements of $W$ in descending order and obtaining $\varphi_{n_1} \ge \varphi_{n_2} \ge \cdots \ge \varphi_{n_N}$ and denoting the corresponding nodes as $O_{n_1}, O_{n_2}, \cdots, O_{n_N}$.

Step 2: Letting $\text{node}_1 = O_{n_1}$, $I = 2$, $J = 2$.

Step 3: Taking out $O_{n_J}$, if for all $1 \le i < I$, the distance between $O_{n_J}$ and $\text{node}_i$ is greater than $\tau$, namely $d(O_{n_J}, \text{node}_i) > \tau$, then $\text{node}_I = O_{n_J}$, going to Step 4; otherwise going to Step 5.

Step 4: $I = I + 1$, $J = J + 1$, if $I > j$ going to Step 6; otherwise, going to Step 3.
Step 5: $J = J + 1$, going to Step 3.
Step 6: Termination.

Algorithm 7.5 does not set the value of $\tau$. Ideally, it should be smaller than the source spacing. However, as the source number and locations are unknown, we may try different values of $\tau$. Obviously, the selected nodes $\{\text{node}_i\}_{i=1}^{k}$ are more likely located close to the sources. Hence we apply the methods discussed in Section 7.1 to estimate the source locations and the energy strengths $\psi_k = \{(\zeta_i^{\mathrm{T}}, U_i)\}_{i=1}^{k}$. Estimating the source number merely with the measurements of $\{\text{node}_i\}_{i=1}^{k}$ and their neighbors is helpful to improve the estimate accuracy and reduce the computation cost, since $k$ is smaller than $N$.

### Estimation procedure based on minimum description length criterion

Suppose the upper bound of source number is $\bar{K}$. Given $k(1 \le k \le \bar{K})$, the maximum likelihood estimate of locations and energy strengths of these $k$ sources, $\psi_k^*$, can be obtained. Then we calculate $F_k$ and MDL($k$) with Eq. (7.14) and Eq. (7.15), respectively. Finally, we get MDL($k^*$) which is the smallest one among MDL($k$), $1 \le j \le \bar{K}$. The procedure is as follows:

$$
\begin{array}{ccccc}
k = 1 & \longrightarrow & \psi_1^* & \longrightarrow & \mathrm{MDL}(1) \\
k = 2 & \longrightarrow & \psi_2^* & \longrightarrow & \mathrm{MDL}(2) \\
\vdots & & \vdots & & \vdots \\
k = \bar{K} & \longrightarrow & \psi_{\bar{K}}^* & \longrightarrow & \mathrm{MDL}(\bar{K})
\end{array}
\Longrightarrow k^* = \arg \min_{1 \le k \le \bar{K}} \mathrm{MDL}(k).
$$

In the end, $k^*$ serves as the estimate of the source number, namely $\hat{K} = k^*$.

### 7.2.3 Simulation and analysis

We study the performance of these two estimation methods in different scenarios of node density and source spacing. The estimate accuracy is defined as the number of correct estimate over the total number of estimates.

In simulation experiments, there are $N$ sensor nodes randomly distributed in an area of $100 \times 100$ square meters in a plane, where three sources are located at $(25, 25)$, $(50, 75)$ and $(75, 25)$, respectively. The energy strength of each source is randomly generated in the range of $[3000, 5000]$. For node $O_i$, its gain factor $g_i = 1$ $(1 \le i \le N)$, and its energy variation is a Gaussian random variable $N(1, 1/50)$. In addition, for an estimation based on the minimum description length, the upper bound of the source number is set to $\bar{K} = 10$.

Fig. 7.10 shows the estimate accuracy corresponding to the number of nodes in the network. With a fixed node number, we have 500 independent runs. The nodes are randomly redistributed in each run. The noise level is $\sigma = 1$ for all nodes.

Fig. 7.10 shows that both methods are able to identify $K$ with high accuracy. The accuracy with the minimum description length criterion exceeds 90% for all values of

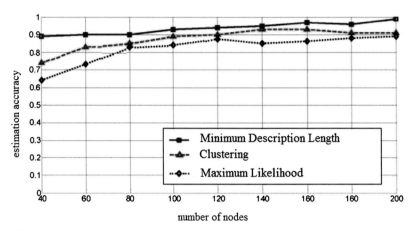

**FIGURE 7.10**

Estimation accuracy ($K = 3$).

$N$ and outperforms the estimate based on node clustering, which performs well when $N > 60$. When $N$ is small, the number of nodes to be selected will be too limited to have a correct clustering. This is why the estimate based on clustering performs worse when $N < 60$. To illustrate the necessity of the penalty term in the minimum description length criterion, the estimate accuracy by the maximum likelihood estimation is also shown in Fig. 7.10 for comparison.

When sources are located close to each other, the measurement model for multisources approximately reduces to the model for a single source. Hence the source spacing will affect the estimate accuracy. As shown in Fig. 7.11, we have $N = 100$ nodes randomly and uniformly deployed in the area of $100 \times 100$ square meters and let two sources be located at $(50 - d/2, 50)$ and $(50 + d/2, 50)$, where $d$ is their distance. Let $d$ increase from 5 to 60. Given the value of $d$, we have 500 independent simulation runs and compute the average estimate accuracy. Fig. 7.12 shows that both of the two methods perform worse when $d$ is smaller than 20. The reason is that the selected nodes are not separated distinctively so that the assumption that the measurement of a node is attributed to only one source may not apply.

## 7.3 Summary

In this chapter, the energy-based acoustic multisource localization algorithm is studied. When the source number is given, the multisource localization problem can be formulated as a maximum likelihood estimation problem. A simple and efficient localization method is proposed by combining source location initialization with the multiresolution search algorithm. When the source number is unknown, node selection and clustering is implemented to estimate the number. The minimum description

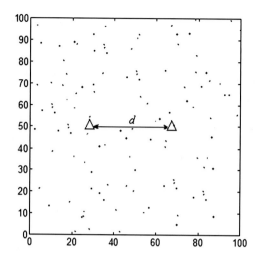

**FIGURE 7.11**

Locations of nodes and sources ($N = 100$, $K = 2$).

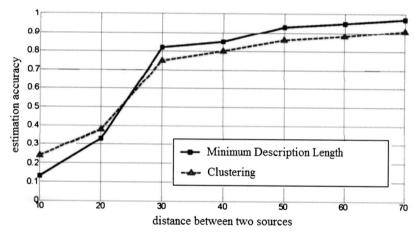

**FIGURE 7.12**

Source spacing vs. estimate accuracy ($N = 100$, $K = 2$).

length criterion is also applied to estimate the source number. Simulation results demonstrate the effectiveness of all algorithms.

# References

[1] H. Akaike, A new look at the statistical model identification, IEEE Trans on Automatic Control 19 (6) (Dec. 1974) 716–723.

[2] M.S. Brandstein, J.E. Adcock, H.F. Silverman, A closed form location estimator for use with room environment microphone arrays, IEEE Transactions on Speech and Audio Processing 5 (1) (Jan. 1997) 45–50.

[3] M. Ester, H.P. Kriegel, A density-based algorithm for discovering clusters in large spatial databases with noise, in: Proceedings of the Second International Conference on Knowledge Discovery and Data Mining, KDD-96, AAAI Press, 1996, pp. 226–231.

[4] J. Ge, X. Chen, The energy-based multisource number estimation in a wireless sensor network, in: 5th International Conference on Information Sciences and Interaction Sciences, ICIS, 2012, pp. 117–122.

[5] S. Haykin, Array Signal Processing, Prentice Hall, Englewood-Cliffs, NJ, 1985.

[6] K.C. Ho, M. Sun, An accurate algebraic closed-form solution for energy-based source localization, IEEE Transactions on Audio, Speech, and Language Processing 15 (8) (Nov. 2007) 2542–2550.

[7] J. Huang, N. Ohnishi, N. Sugie, Sound localization in reverberant environment based on model of the precedence effect, IEEE Transactions on Instrumentation and Measurement 46 (Aug. 1997) 842–846.

[8] K.M. Kaplan, Q. Le, P. Molnar, Maximum likelihood methods for bearings-only target localization, in: Proc. IEEE ICASSP, vol. 5, 2001, pp. 3001–3004.

[9] D. Li, Y. Hu, Energy-based collaborative source localization using acoustic micro sensor array, EURASIP Journal on Applied Signal Processing 2003 (1) (2003) 321–327.

[10] W. Meng, W. Xiao, L. Xie, An efficient EM algorithm for energy-based multi source localization in wireless sensor networks, IEEE Transactions on Instrumentation and Measurement 60 (3) (2011) 1017–1027.

[11] J.K. Nelson, M.U. Hazen, M.R. Gupta, Global optimization for multiple transmitter localization, in: Milcom, Feb. 2006, pp. 1–7.

[12] M. Omologo, P. Svaizer, Acoustic source location in noisy and reverberant environment using CSP analysis, in: Proc. ICASSP, Atlanta, GA, 1996, pp. 921–924.

[13] Y. Oshman, P. Davidson, Optimization of observer trajectories for bearings-only target localization, IEEE Transactions on Aerospace and Electronic 35 (3) (July 1999) 892–902.

[14] C.W. Reed, R. Hudson, K. Yao, Direct joint source localization and propagation speed estimation, in: Proc. ICASSP, Phoenix, AZ, 1999, pp. 1169–1172.

[15] J. Rissanen, Modeling by shortest data description, Automatica 14 (1978) 465–471.

[16] J. Rissanen, Universal coding, information, prediction, and estimation, IEEE Transactions on Information Theory 30 (2) (July 1984) 629–636.

[17] X. Sheng, Y.H. Hu, Maximum likelihood multiple-source localization using acoustic energy measurements with wireless sensor network, IEEE Transactions on Signal Processing 53 (1) (2005) 44–53.

[18] Q. Shi, C. He, A new incremental optimization algorithm for ML-based source localization in sensor networks, IEEE Signal Processing Letters 15 (1) (2008) 45–48.

[19] J.O. Smith, J.S. Abel, Closed form least square source location estimation from range difference measurements, IEEE Transactions on Acoustics, Speech, and Signal Processing 35 (12) (Dec. 1987) 1661–1669.

[20] K. Suyama, K. Takahashi, R. Hirabayashi, A robust technique for sound source localization in consideration of room capacity, in: Proc. IEEE Workshop Appl. Signal Processing Audio Acoust., New Paltz, NY, 2001, pp. 63–66.

[21] L.G. Taff, Target localization from bearings-only observations, IEEE Transactions on Aerospace and Electronic 3 (1) (Jan. 1997) 2–10.

[22] H. Wang, P. Chu, Voice source localization for automatic camera pointing system in video conferencing, in: Proc. ICASSP, vol. 1, 1997, pp. 187–190.

[23] M. Wax, T. Kailath, Detection of signals by information theoretic criteria, IEEE Transactions on Acoustics, Speech, and Signal Processing 33 (1) (April 1985) 387–392.

[24] K. Yao, R.E. Hudson, C.W. Reed, D. Chen, F. Lorenzelli, Blind beam forming on a randomly distributed sensor array system, IEEE Journal on Selected Areas in Communications 16 (October 1998) 1555–1567.

# Index

Printed in the United States
By Bookmasters